Prespacetir

Volume 2 Issue 13
December 2011

Hints of Higgs Boson at 125 GeV Are Found: Congratulations to All the People at LHC!

Chief Editor:
Huping Hu, Ph.D., J.D.

Editor-at-Large:
Philip E. Gibbs, Ph.D.

Advisory Board

Dainis Zeps, Ph.D., Senior Researcher, Institute of Math. & Computer Sci., Univ. of Latvia, Latvia
Matti Pitkänen, Ph.D., Independent Researcher, Finland
Arkadiusz Jadczyk, Professor (guest), Center CAIROS, IMT, Univ. Paul Sabatier, Toulouse, France
Philip E. Gibbs, Ph.D., Independent Researcher, United Kingdom
Jonathan J. Dickau, Independent Researcher, United States
Stephen P. Smith, Ph.D., Visiting Scientist, Physics Dept., UC Davis, United States
Lawrence B. Crowell, Ph.D., Alpha Institute of Advanced Study, Hungary
Andrei Khrennikov, Professor, In'tl Center for for Mathematical Modeling, Linnaeus Univ., Sweden
Elemer E. Rosinger, Emeritus Prof, Dept. of Math. & Applied Math., Univ. of Pretoria, South Africa

ISSN: 2153-8301   Prespacetime Journal   www.prespacetime.com
Published by QuantumDream, Inc.

Table of Contents

Special Reports

Refined Higgs Rumours
Philip E. Gibbs — 01-02

Higgs Boson Live Blog: Analysis of the CERN Announcement
Philip E. Gibbs — 03-25

Articles

Has CERN Found the God Particle? A Calculation
Philip E. Gibbs — 26-34

Electron Spin Precession for the Time Fractional Pauli Equation
Hosein Nasrolahpour — 35-41

GR Articles

Plane Wave Solutions of Weakened Field Equations in a Plane Symmetric Space-time-II
Sanjay R. Bhoyar & A. G. Deshmukh — 42-52

Plane Wave Solutions of Field Equations of Israel and Trollope's Unified Field Theory in V_5
Gowardhan P. Urkude, J. K. Jumale & K. D. Thengane — 53-64

Essays

If the LHC Particle Is Real, What Is One of the Other Possibilities than the Higgs Boson?
Huping Hu & Maoxin Wu — 65-67

What is Reality in a Holographic World?
James Kowall — 68-170

News

Searching for Earth's Twin
Phillip E. Gibbs — 171-173

Special Report

Refined Higgs Rumours

Philip E. Gibbs[*]

Abstract

We report here refined Higgs rumours before the December 13, 2011 announcement by CERN about Higgs search results of LCH.

Key Words: Higgs Boson, rumour, CERN, LHC.

December 12, 2011: Refined Higgs Rumours

Jester has kindly provided some more refined rumours to give us something to talk about and make the time go quickly while we wait for the Big Event. Here are my comments

"The Standard Model Higgs boson is excluded down to approximately 130 GeV, but not below."

Very nice but this will be using the WW channel. I don't fully trust this decay mode for exclusions in the lower energy range because of the poor energy resolution. Previously we have seen both exclusions and excesses near this region. It could mean that there is a non-standard Higgs Boson at 140 GeV that might appear to have lower signal because e.g. it decays to something unknown. It could also just be an effect of the poor WW resolution. I will be looking to see what happens at the 140 GeV point in the combined diphoton and ZZ -> 4l channel without WW to understand this better.

"As already reported widely on blogs, both experiments have an excess of events consistent with the Higgs particle of mass around 125 GeV."

The interesting thing here is going to be to see how big the excess is when the two experiments are combined. Combining the excess strengths is not just a matter of adding in quadrature. That gives just a crude approximation. I will do a better approximation when I have the data. I am also wondering whether the size of the signal is consistent with a Standard Higgs or bigger. I think it has to be bigger by a factor of two because we only expect 2-sigma significance without the WW channel. I will also look forward to seeing how this shows up on the raw event count plots. Overall a lot of what is seen here will be noise because the sensitivity is still relatively low, but a high sigma combined excess would mean there is probably something.

"The excess is larger at ATLAS, where it is driven by the $H \rightarrow \gamma\gamma$ channel, and supported by 3 events reconstructed in the $H \rightarrow ZZ^* \rightarrow 4l$ channel at that mass. The combined significance is around 3 sigma, the precise number depending on statistical methods used, in particular on how one includes the look-elsewhere-effect."

---

[*] Correspondence: Philip E. Gibbs, Ph.D., Independent Researcher, UK. E-Mail: phil@royalgenes.com
Note: This report is adopted from http://blog.vixra.org/2011/12/12/refined-higgs-rumours/

How close in energy are these three events? That could be key. In any case we should not expect much contribution from ZZ at 125 GeV yet. The channel is just not sensitive enough with 10/fb and will be mostly weighted out in the combination with diphoton.

"CMS has a smaller excess at 125 GeV, mainly in the H→γγ channel, but their excess in H→4l is oddly shifted to somewhat lower masses of order 119 GeV. All in all, the significance at 125 GeV in CMS is only around 2 sigma."

No surprise that the CMS ZZ result is inconsistent. There is too much noise in this channel at < 130 GeV to know what is the real signal at this point. At the end of next year it will start to come through. For now it will add just a little contribution to the diphoton channel. 2 sigma is very little but when combined with ATLAS it adds up.

"With some good faith, one could cherish other 2-sigmish bumps in the γγ channel, notably around 140 GeV. Those definitely cannot be the signal of the Standard Model Higgs, but could well be due to Higgs-like particles in various extensions of the Standard Model."

Indeed, but the big question is whether the 140 GeV bumps previously seen in the ZZ channel are still there. This is now very sensitive at 140 GeV so we should know something. Since there is no rumour about this it might mean that nothing is there and the diphoton bump is just the remainder of the big excess seen there in the summer.

Aside from all that we are interested to see what remains at higher mass, especially around 240 GeV and 600 GeV. Stay tuned.

References

1. http://blog.vixra.org/2011/12/12/refined-higgs-rumours/

Special Report

Higgs Boson Live Blog: Analysis of the CERN Announcement

Philip E. Gibbs[*]

Abstract

We report here live on LHC Higgs results announcement by CERN on December 13, 2011. The result is very convincing if one starts from the assumption that there should be a Higgs Boson somewhere in the range. Everywhere is ruled out except 115 GeV to 130 GeV and within that window there is a signal with the right strength at around 125 GeV with 3 sigma significance. CERN will have to wait for that to reach 5 sigma to claim discovery and next year's data should be enough to get there or almost. I calculate that they will need 25/fb per experiment at 7 TeV to make the discovery. A big congratulation goes to everyone from the LHC, ATLAS and CMS who found the clear hints of Higgs when it hid in the hardest place.

Key Words: Higgs Boson, announcement, CERN, LHC, ATLAS, CMS.

December 13, 2011: Higgs Boson Live Blog: Analysis of the CERN Announcement

Good morning and welcome to what is expected to be an exceptional day for physics as CERN announces important new results in their hunt for the elusive Higgs Boson. Here in one mammoth expanding post I will be reporting on the search for the Higgs Boson in straight forward terms free form silly analogies and patronizing phrases such as "for the layman". I hope that many interested people with varying degrees of foreknowledge will find the level helpful. I will explain the basic preliminaries first but if there is anything you don't understand just Google it or wing it.

The present excitement started to build during the summer when it became clear that the Large Hadron Collider experiment was gathering data at a much higher rate than anticipated, meaning that they would soon be able to tell whether the Higgs boson exists or not and most importantly, what mass it has.

I am a theoretical particle physicist based near London independent of the teams working at CERN, and I have been following events at the Large Hadron Collider and blogging about them since it started colliding protons in 2009. In a minute I will answer a few basic questions about the Higgs for the uninitiated, including the Paxman question "What does the Higgs boson look like?" Then I will be live-blogging the events from CERN as they happen, so first let's look at the schedule of today's events.

- **14:00** – Fabiola Gianotti, spokesperson for the ATLAS collaboration delivers a 30 minute summary of their latest developments. ATLAS is the largest particle detector

---

[*] Correspondence: Philip E. Gibbs, Ph.D., Independent Researcher, UK. E-Mail: phil@royalgenes.com
Note: This report is adopted from http://blog.vixra.org/2011/12/13/the-higgs-boson-live-from-cern/

ever built and it sits on an intersection point of the Large Hadron Collider rings to observe the trillions of particle collision events taking place.

- **14:30** – Guido Tonelli will talk about similar observations at the CMS experiment. CMS is another equally sophisticated but different and complementary detector placed diametrically opposite ATLAS on the LHC ring gathering another independent set of collision data.
- **15:00** – When the talks end, which may not be on time, there will be an hour long technical discussion between the scientists about each others results. Until these talks the two 3000 strong teams of physicists had not officially compared their data so there will be much to talk about.
- **15:20** – At this time we expect a release of information and pictures to the press as the scientific discussion continues.
- **16:30** – Press conference. Questions and Answers from the experts

During these events I will be posting news and exclusive analysis right here as it happens. You can refresh this page for updates and post your own views and observations in the comments section. However, please accept that I may delete comments that I consider unhelpful to a general audience. You can continue to post broader material on the previous post about the rumours

Amongst other things I will be attempting to combine the results in real time as soon as the necessary plots become available. The CERN director General has forewarned us that the announcement today will not provide conclusive evidence for the existence, or non-existence of the Higgs boson, but that could be because the two experiments have not had time to combine their results. The official combination will not be ready until next year because the full computational process is long and difficult. However, it is possible to do a quick approximate "bloggers" combination that will allow us to anticipate what the eventual result will look like. In fact the method has been shown to be reasonably accurate in the past. I will be doing more combinations right here today.

Let me just reiterate that again. *My combinations are approximate*. They assume a flat normal probability distribution. That is a good approximation that improves as more data is added. They also assume that there are no correlations between uncertainties among the different parts of the experiments. This is not the case. Such correlations have a small effect that does not diminish with more data. In order to claim a discovery using a combination the collaborations will have to get together and do an official version the hard way and that will take time. However, my quick combination method is good enough to give a very good idea of what the final result will look like and it is certainly not "Nonsense" as some of the experimenters have tried to claim.

Why is the Higgs Boson so special?

During the 1960s and 1970s theoretical physicists using data from the first generation of particle accelerators assembled a theory of elementary particles known as the Standard Model. It included familiar particles such as the electron, photon and neutrinos as well as unseen quarks that bind to form protons and neutrons inside the atom. All the particles in the standard model are of two types with one exception.

The particles which build up matter are all *spin-half fermions* which obey an equation formulated by Dirac in 1928. This includes the three generation of quark pairs and the three corresponding pairs of leptons, the electron, muon and tauon with their neutrino partners. Each of these has an antimatter partner so there are 24 distinct fermions in the Standard Model. The second set of particles are the *spin-one bosons*. These play the role of binding together the fermions with the electromagnetic force (the photon) the strong nuclear force (the gluons) and the weak nuclear force (the W and Z bosons) Of these only the W is charged and so has a distinct anti-particle, meaning that there are 5 different bosons.

Aside from these it was found that the standard model required one further particle. It was known that a consistent model of spin-half fermions and spin-one bosons free from infinities required gauge symmetry, that is a mechanism that would in theory make the bosons massless. On the other hand, nature had shown that the bosons that mediate the weak nuclear force must have mass. The solution was a mechanism worked out around 1960 by a number of physicists that introduces an unusual field into the theory. The field has an unorthodox energy potential that is minimised away from the central point of symmetry so that the value of the field in the vacuum state of space-time must be shifted away from the central point, thus breaking the underlying symmetry and giving mass to some of the particles.

Peter Higgs, one of the pioneers of this mechanism, pointed out that the remnant of this field in its broken form would have excitations corresponding to a unique elementary particle that might be observed as final confirmation of the theory. Unlike the other particles in the Standard Model, this one would be a *spin-zero boson*. Observation of this hypothetical particle named the Higgs Boson in his honour is what the Large Hadron Collider has been looking for 50 years later.

What does the Higgs Boson look like?

The Higgs Boson exists only for fleeting moments as a fuzzy quantum wave on scales smaller than the inner workings on the proton. It is therefore impossible both theoretically and practically to "see" it in the normal sense of the word. What we can see are traces of its existence in data gathered from countless collisions between high energy protons in the Large Hadron Collider.

In the LHC at CERN on the Swiss-Franco border near Geneva, physicists have been accelerating protons to unprecedented high energies in a circular underground ring 27 km in circumference. When the protons are brought together in a head-on collision the energy can form new particles, perhaps including some never observed before such as the Higgs boson. Many trillions of collisions have been observed but the processes that form a Higgs boson are so rare that only a few thousand are likely to have been created in the experiment so far.

Once created a Higgs boson should live for a fleeting 10^{-22} seconds, enough time for it to travel between 10 and a hundred times the width of the protons from which it emerged. Then it decays, usually into other particles, most often a matched pair of bottom/anti-bottom quarks which have a much longer lifetime of 10^{-12} seconds. As the bottom quarks fly apart a string of gluon flux stretches between them before breaking to form new quarks. These emerge along with the decay products of the bottom quarks as jets of hadrons that reach the detectors. Sometimes the bottom quarks will each decay into another quark plus a lepton (electron or muon) with an accompanying neutrino. The lepton makes tale-tale tracks in the detector while the neutrino flies off without a trail only to be guessed at when they add up the energy of all the other particles and notice that some is missing. Unfortunately there are many other less remarkable processes that produce similar jets and leptons at the LHC making it very difficult to observe the Higgs Boson when it decays in this way.

If the latest rumours about the measurements at CERN are correct the Higgs Boson could have a mass approximately equal to that of a Caesium atom. If this is correct about one in 500 of the Higgs bosons produced will decay into two high energy photons that fly away in opposite directions. Unlike the bottom quarks these fly away cleanly carrying all their energy and momentum to the inner layers of the detectors where a surrounding vessel of liquid argon has been placed to capture them. There they produce a shower of lower energy particles that are carefully tracked so that their energy and trajectory can be measured to reveal the parameters of the original photon. During all of this years run at the LHC this may have happened only a dozen times in each detector, but it could be enough to reveal the Higgs Boson.

Such photons will be thousands of times more energetic than the harmful gamma rays that emanate from nuclear reactions, but they are still photons identical to those of light which differ only by having less energy. If you want to know what the Higgs boson looks like it is the faint glow of these rare photons that answers the question most directly. In the LHC they shine faintly among the brighter radiation of other processes that produce equally energetic gamma rays. The ones coming from the Higgs Boson can only be noticed when enough have been detected to show up as a slightly brighter peak in the energy spectrum of thousands of observations. It is this that we are hoping to hear news of today.

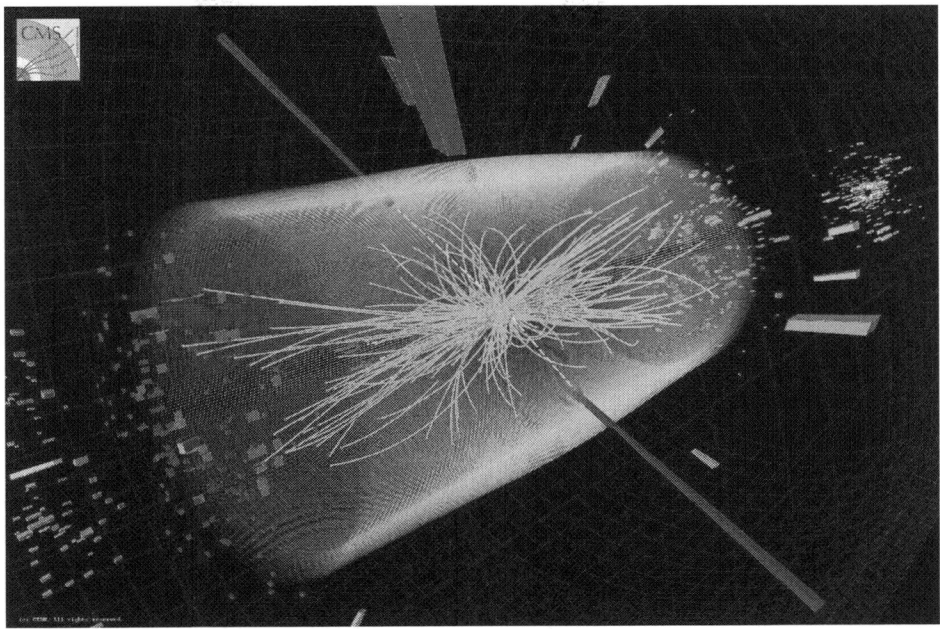

A typical event with two high energy photons as recorded in CMS

Will the LHC find the Higgs Boson?

The theory of the Higgs Boson has been around a long time and all the other particles of the standard model have been found. Several of them were found after they were predicted by the model, especially the gluons, W and Z bosons and top quark. This means that the theory of the Standard Model is in very good stead experimentally. Indeed, physicists have been hoping for some experimental deviation from its predictions for decades and have come away disappointed. Every experiment just seems to confirm its correctness with ever more accuracy. (There are some exceptions such as measurement of the muon magnetic anomaly and the cosmological observation of dark matter that seem to point to something beyond the standard model at higher energy)

With such success it is no wonder that the theorists are quite confident that the Higgs Boson will be found as the last missing piece of the Standard Model. However, experiment is the ultimate judge of nature and theorists are not always right. A minority of physicists notably including Stephen Hawking and Nobel laureate Martinus Veltman have said that they do not believe the Higgs Boson will be found because according to their theories it cannot exist. They are considered contrarians by other physicists but until the "Goddamned" particle has been found nobody can be certain.

One thing that is sure is that the Large Hadron Collider will either discover the Higgs Boson or rule it out as predicted by the Standard Model. If all goes well this will be achieved before the end of 2012, perhaps much sooner. It has been said that if the Standard Model Higgs Boson is ruled out it will be an even greater discovery than its mere existence. This is not just excuses for what some people may portray as a failure. Such a result would indeed by a breakthrough inevitably leading to a new and better understanding of physics.

It is also possible that the Higgs Boson exists but that its characteristics are different from those of the Standard Model. In particlular, it may decay into other lighter unknown particles making it hard to detect. In that case it might appear not to be there even though it is. That will still count as ruling out the Standard Model Higgs but until further experiments are done it will not be known whether it does not exist at all, or is merely hidden from view by non-standard processes. Another even more exciting possibility is that there is more than one Higgs Boson possibly including some heavier versions that are charged. This is predicted by some grander theories such as supersymmetry

However, results from the LHC so far suggest that whatever happens there will be something positive to report today. It will not be quite a full discovery but it will be a strong signal that something like a Higgs Boson exists. Although we have heard some quite detailed rumours already, it is only by seeing the actual graphs that we can get a good idea of what the possibilities are. All physicists are now eagerly waiting to see them.

What will we learn?

You might think that since the Higgs Boson was predicted 50 years ago its discovery today will not be very exciting news. Indeed, before the LHC started collecting data, many physicists saw its discovery as inevitable and uninteresting. This view has changed, partly because nothing else has been quick to manifest itself at the LHC as hoped. This means that the Higgs Boson is likely to be the leading discovery of any new physics.

The mass of the Higgs Boson is a free parameter in the Standard Model. Once it is known, all other features such as its lifetime and interaction rates can be calculated. However, analysis of the physics of the Standard Model shows that if the mass is not within strict limits the theory will break down at higher energies. In particular, if it is too light the vacuum will not be sufficiently stable, but we know that this cannot be happening in the real world. The mass range left where the Higgs Boson can still be found includes a range where this would be a problem for the theory.

If it is lighter than 126 GeV then that may be an indication of new physics that could be found with more data. The theory of supersymmetry which is very popular with theorists actually favours the lighter Higgs and corrects problems with the stability of the vacuum, but it does not support well a heavier mass. For these reasons today's announcement could signal the directions of research for future physics depending on what mass is indicated by the experiments.

What will we be looking out for today?

Despite the rumours, it is not certain exactly what will be shown today, but we are hoping for full reporting of all the results in the Higgs search from the two individual experiments. This would include the analysis of each possible decay mode that the experiments can currently observe plus two combination of results from all channels, one for ATLAS and one for CMS. The amount of data collected this year corresponds to an integrated luminosity of 5 inverse femtobarns (5/fb) in each experiment so anything less than this is not complete.

There are three sets of decay channels that are currently of special relevance to the search,

- **diphoton** (a.k.a. digamma) where the Higgs Boson decays directly to two photons
- **WW -> lvlv** where the Higgs Boson decays to two oppositely charged W bosons which then decay to electrons or muons and associated neutrinos
- **ZZ -> 4l** where it decays to two neutral Z bosons that then each decay to two oppositely charged electons or muons making four leptons in total.

If recent rumours are correct it is the diphoton channel that holds the most interest with a signal of a possible Higgs Boson at a mass of 125 GeV, but we will be very interested in the other channels to see if there is any supporting evidence or signs of anything at other masses. It will be especially interesting to see of the earlier weak signal at 140 GeV has gone away entirely. These and other channels may provide signs of something interesting at higher masses but most likely there will just be a strengthening of the evidence for exclusion above 140 GeV.

What do the plots mean?

During the presentations delivered by the collaborations today we will see a lot of new graphs. If you are not familiar with these they will require some explanation. The ones that everyone will be looking out for are the "Brazil band" plots, named for their distinctive green and yellow bands. These plots are the main way of showing the results from each Higgs Boson decay mode as well as the all important combinations.

Here is the best LHC combination plot for Higgs boson searches made public prior to today. It incorporates about a third as much data as gathered during the whole year and was shown in November at the Hadron Collider Physics conference, but I have redrawn it to add some extra features. (With any plot on this blog you can click on the image to enlarge for a clearer picture)

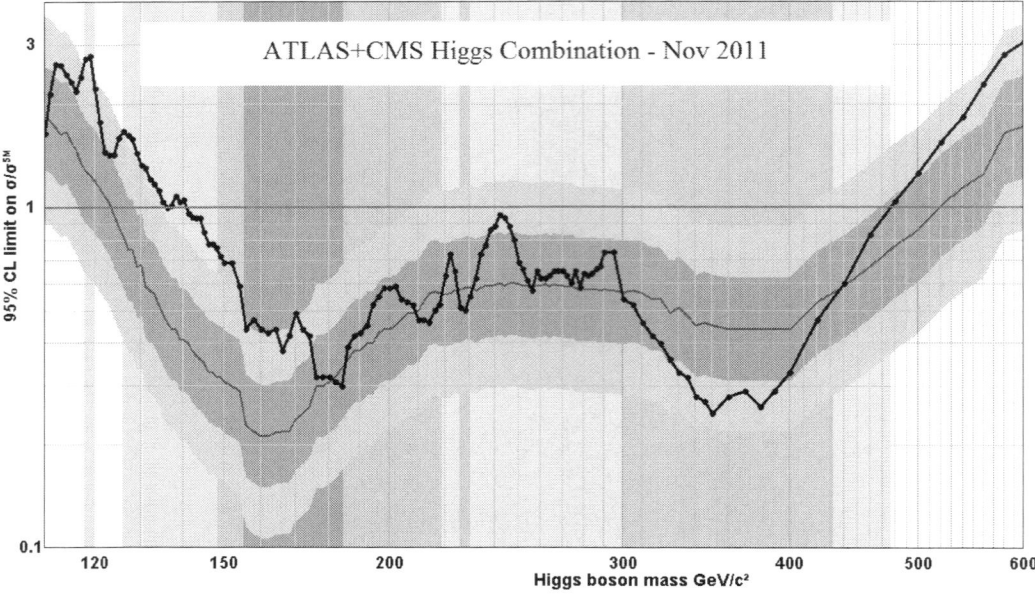

The horizontal axis is marked with the range of possible masses for the Higgs Boson. The units are Giga electron-Volts as an energy equivalent of mass. This is the standard way to measure mass in an accelerator experiment. If the Higgs Boson has a mass of 125 GeV as rumoured you should be able to see where it would appear on this plot.

The black line is usually called "Observed CL_s" and represents the calculated result from all the experiments. Its value for any given mass gives a quantity labelled "95% Confidence Level limit for σ/σ^{SM}" on the vertical axis. What does this mean exactly? Take an example; At 200 GeV the observed CL_s has a value of about 0.6. What this says is that if the signal cross-section over all the decay modes were just 0.6 times the amount expected if the Standard Model is correct and the Higgs Boson has a mass of 200 GeV, then there would be a 95% probability of seeing more events than they did. This is a roundabout way of saying that we have seen far too few events, so we can rule out the Higgs Boson at this mass with some confidence.

When the black line descends below the red horizontal line at 1.0 on the vertical axis, people sometimes say that the Higgs Boson has been ruled out at 95% confidence level at this mass. This is not strictly correct because such confidence would depend on our prior assessment of the probability for the existence of the Higgs Boson in this mass range in the first place, and also the "Look Elsewhere Effect" would have to be considered. Such knowledge is subjective and dependent on outside influences, but loosely thinking you can interpret it that way.

In the background of the plot I have shaded areas in various grades of pink. The lightest pink indicates an exclusions at 95% confidence. This is often stated as 2-sigma significance because statistically it corresponds to 2 standard deviations away from the normal expectation. Darker shades of pink indicate 3-sigma and 4-sigma confidence. Until recently it was generally accepted that 2-sigmas was enough to rule out the presence of the Higgs Boson at a given mass, but recently people have said they want 5-sigma significance, the same as for the discovery of a new particle. I think in reality most people will accept 3-sigma for exclusions.

But we are no longer just interested in exclusions. How do we know from this plot if the Higgs Boson has been seen? This is where the yellow and green bands come in. The central blue line indicates the expected value under the condition that no Higgs Boson exists at a given mass. The green and yellow bands are the 1-sigma and 2-sigma deviations from that expectation. This means that if there is no Higgs Boson the observed CLs line should wander within these two bands. Statistically it is likely to go outside the yellow bands for about 5% of its range. When we look at the plot we see that this is indeed the case. Despite the excess exceeding 2-sigmas around the 140 GeV region we can only say that the result is consistent with the lack of a Higgs Boson over the whole range. That is not a very encouraging way to put it. Notice that mass ranges where there are excesses will be background shaded in grades of green.

Can we at least say that the plot is also consistent with the hypothesis that there is a Higgs Boson somewhere in the mass range? We can see that it is excluded over the range from 140 GeV to 480 GeV at 2-sigma significance but we can still accept the possibility that it is in the low or high mass region. there are theoretical reasons to strongly doubt that it is at the high mass end so the range 115 GeV to 140 GeV is the best bet.

It is possible to display the same results in a different way that handles the existence and exclusion of the Higgs Boson in a more symmetrical way. This is sometimes called the "best fit" plot or "signal" plot and for the combination above it would look like this.

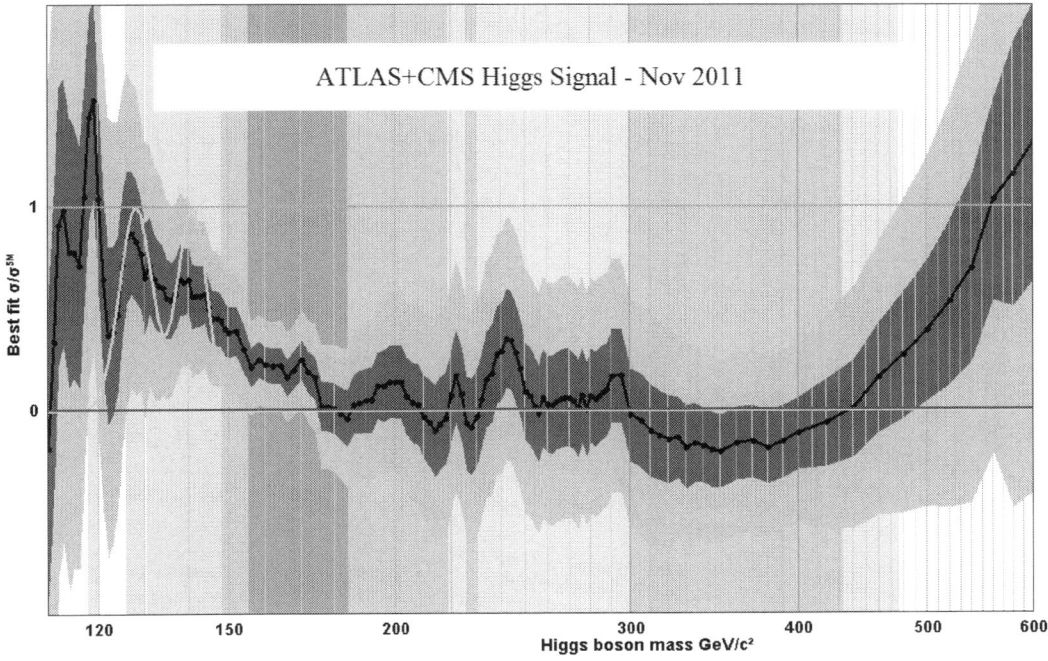

The experimenters don't often display their results this way, but as theorist I find it the best plot to give a feel for where we stand. If I can get the data from the talks today I may show some of these plots.

The black line varies around a range of signal values where a signal of zero would indicate just the Standard Model background with no Higgs Boson and a signal of one is just the right strength for its existence. The blue and cyan bands are error bands (mostly statistical) around the observed data. When the blue and cyan error bands extend over the whole range between the red line at zero and the green line at one we really have no indication either way for a Higgs Boson or its exclusion in the mass range. However, when it starts to settle on one of either the red or green line and moves clear of the other, then we know that we have the right signal strength for the presence or absence of the Higgs Boson.

What will happen after today?

Whatever comes out today there will still be a lot more work to be done. At the moment the LHC is shutdown for the Winter to allow for maintenance and to save electricity at a time when domestic demand is highest. It will startup again in February next year. Meanwhile the physicists will be using the time to continue the analysis of the data already collected during 2011 and that will include preparing the official combination of today's results from ATLAS and CMS.

Next year the LHC will run again, probably at a slightly higher energy of 8 TeV rather than the 7 TeV used this year. It is expected to collect three times as much data in 2012 as it did in 2011 so by the end of the year they will have a total of at least 20/fb on tape for each of ATLAS and CMS. If they don't already have enough data to know whether the Higgs Boson exists they almost certainly will by then.

More importantly, they will start to study the properties of the Higgs Boson to check that it matches the standard model by decaying into all types of lighter particle at the predicted rates. If it doesn't then they will know that there is new physics outside the Standard Model to be understood.

That assumes that the standard Higgs Boson will show up. If it doesn't they will have the job of looking for what replaces it . That can be done by looking at interactions between W bosons which should get stronger with increasing energy if there is no Higgs Boson until something gives. Present rumours suggest that the Higgs does exist but these WW scattering experiments will still be interesting.

After 2012 the LHC will shutdown for about 18 months to prepare it for running at higher energies, probably 13 TeV during 2015 and 14 TeV later. They will be searching for more new particles but they will also checking the parameters of the Standard Model including the Higgs Boson in more detail to eek out any signs of dark matter or anything else not seen before. The LHC will continue to run at higher luminosity and possibly even higher energy for perhaps another 30 years. This is just the beginning of what it has to do.

Live Blog Starts Here

09:00 (times are Central European)

This morning ATLAS have released an update to the Higgs search in the WW -> lvlv channel. They are using 2.05/fb in place of the previous 1.66/fb so it is only a small advance. This had been around for some time unofficially but was not shown at the HCP2011 conference, Hopefully it will be obsolete in a matter of hours but here is the plot anyway. It provides 95% exclusion from 145 GeV to 200 GeV.

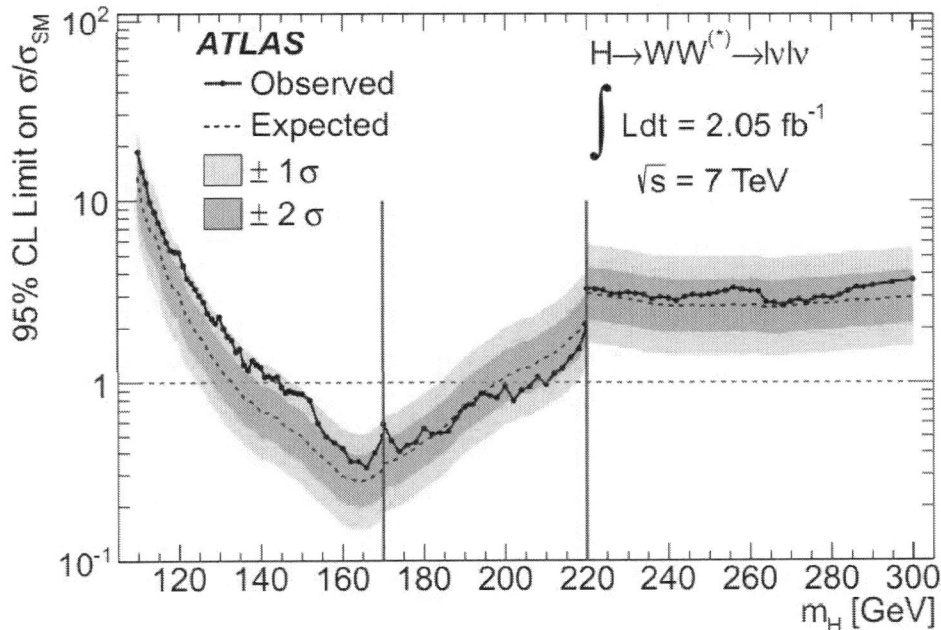

11:45

Just to remind everyone, the official build-up for this event is as follows: "These results will be based on the analysis of considerably more data than those presented at the summer conferences, sufficient to make significant progress in the search for the Higgs boson, but not enough to make any conclusive statement on the existence or non-existence of the Higgs."

If you come here expecting a life-changing discovery to be announced you will be disappointed, but if you want to see some science in action taking a small step forwards you may enjoy.

12:00

With two hours to go the auditaurium was already full.

13:47

Here in the UK the BBC are already running reports on the network news. They are saying that each experiment is finding a blip in the same place giving a strong hint of the Higgs.

14:00

Speakers introduced, talks getting underway

14:15

ATLAS have updated the three most sensitive channels diphoton to 4.9/fb ZZ->4l to 4.8/fb and WW->lvlv to 2.1 (as above)

14:25

I have the CMS Combo, here it is with exclusion from 130 GeV up. Excess seen at about 123 GeV of 2.5 Sigma

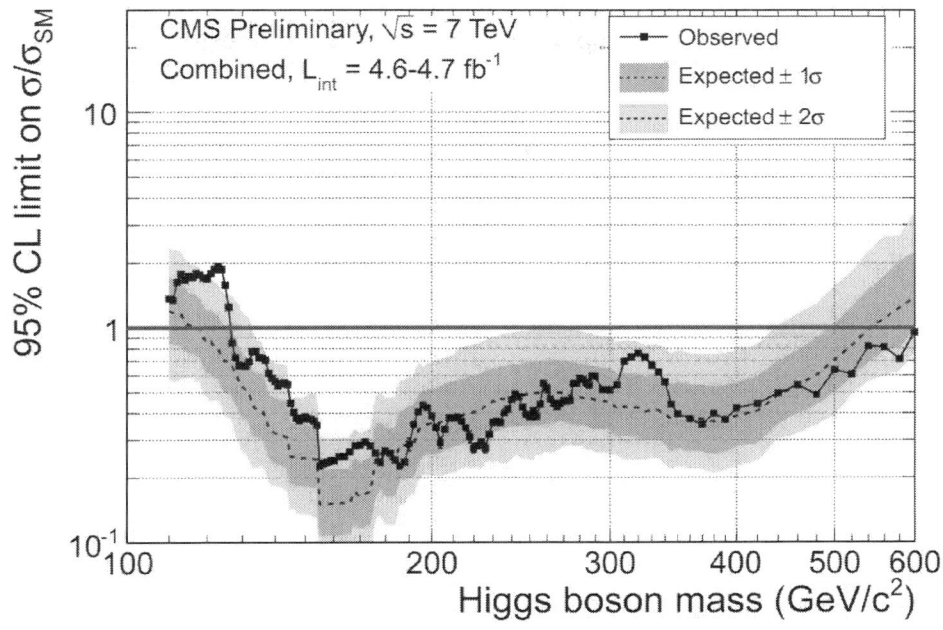

14:30

Here is the CMS diphoton plot shwoing where the excess comes from, but there are other excesses nearly as big

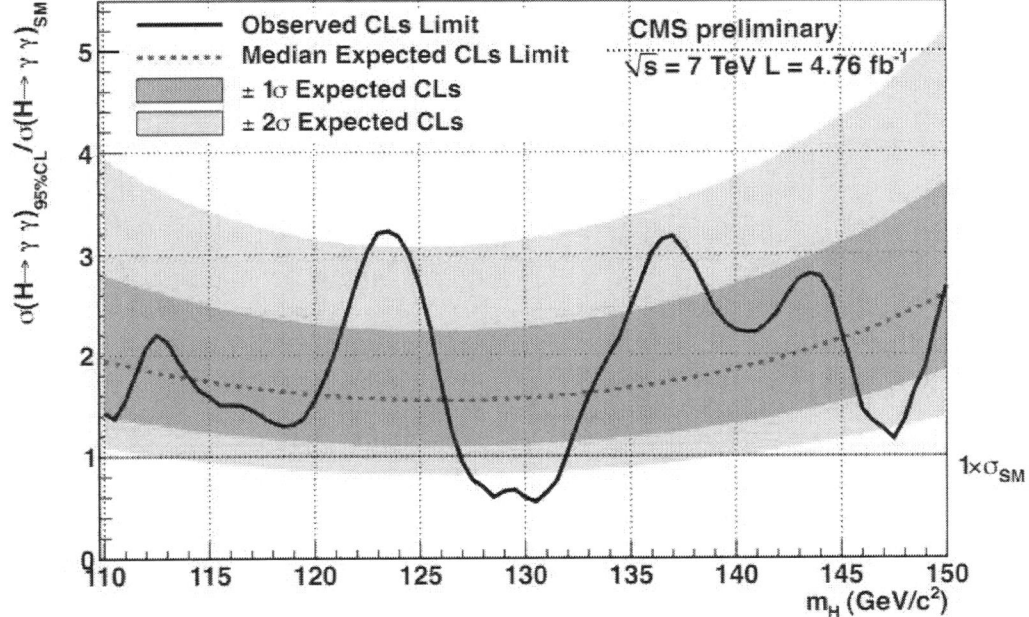

14:32

Here is the ATLAS version from the talk. Updated from conference notes.

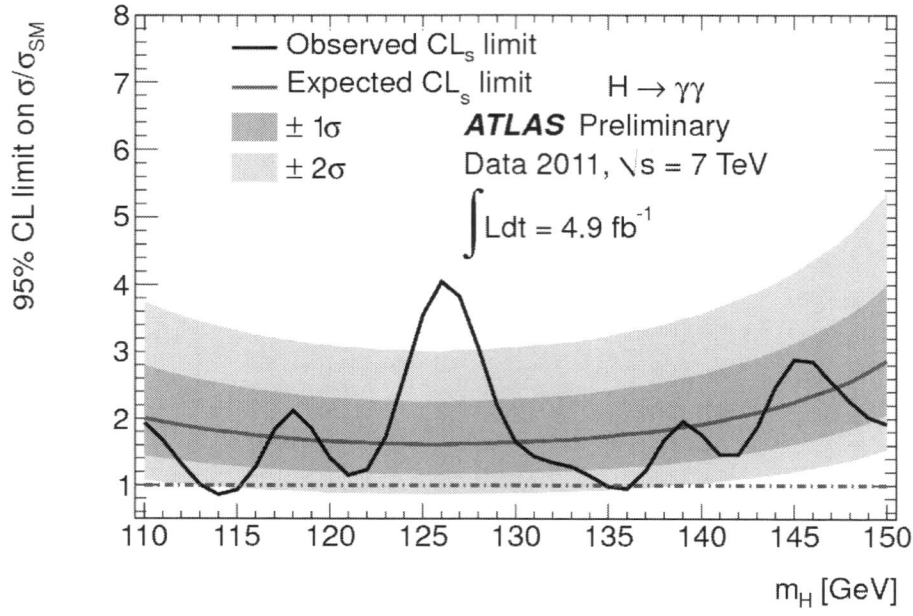

14:36

The CMS ZZ->4l clearly rules out the 140 GeV possibility, but has an excess at lower mass.

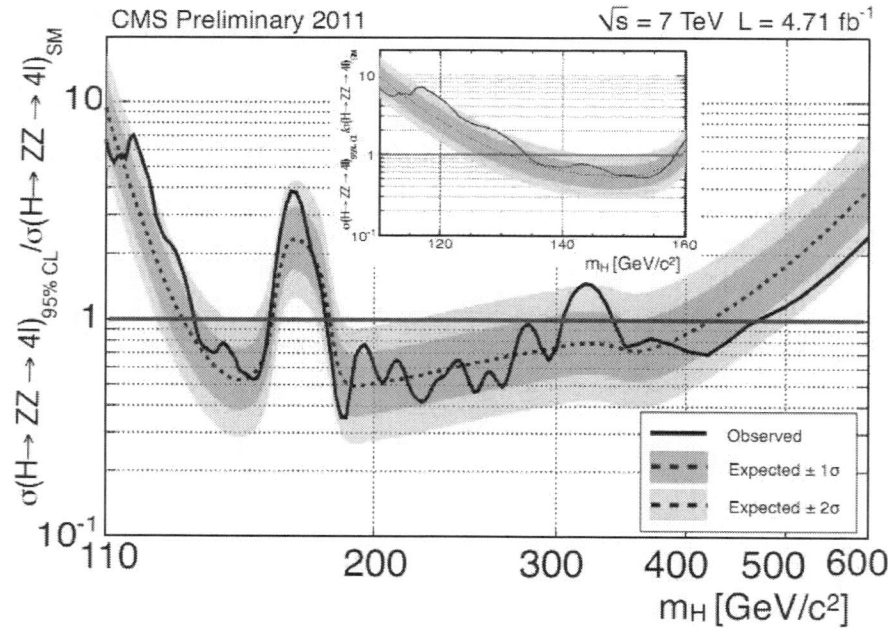

14:43

ATLAS ZZ->4l and full combo from the talk. Updated from conference notes.

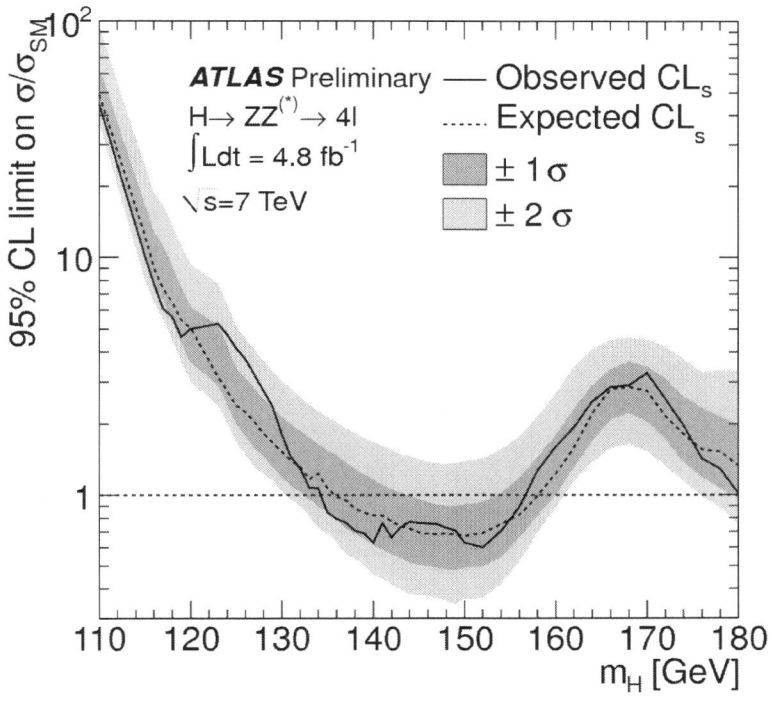

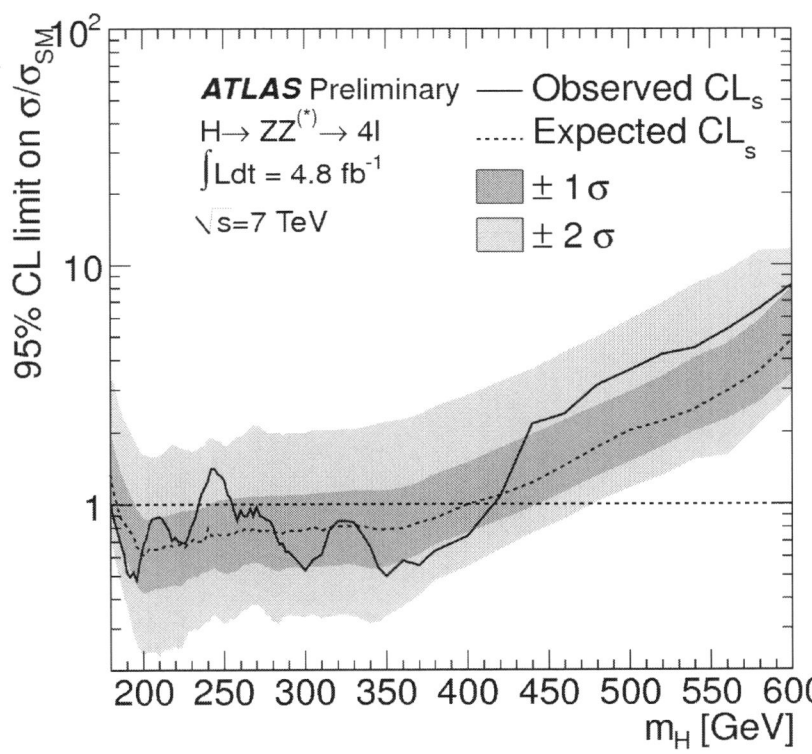

ATLAS full combo from the talk. Updated from conference notes.

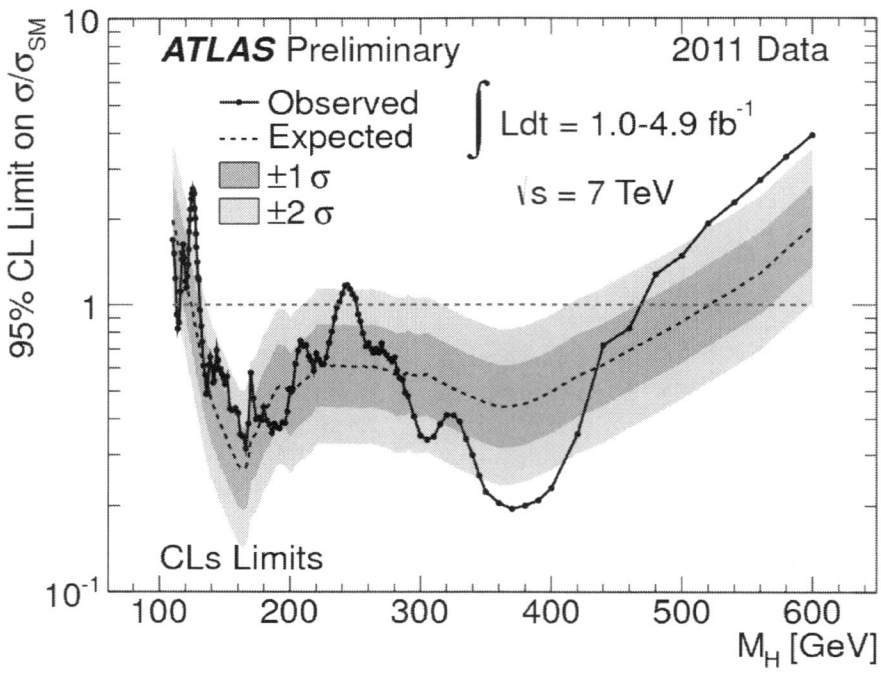

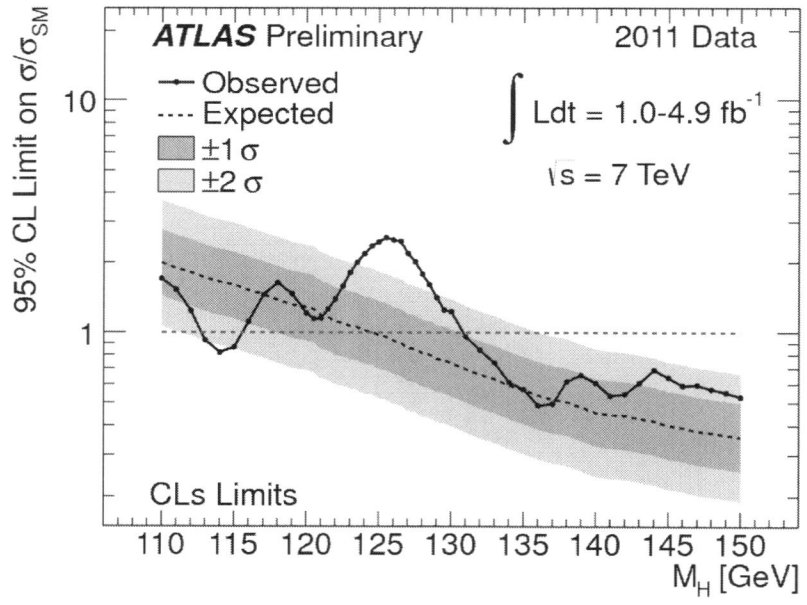

14:49
First talk is over, now over to CMS
CMS have two versions of the WW channel, cutbased and BDT

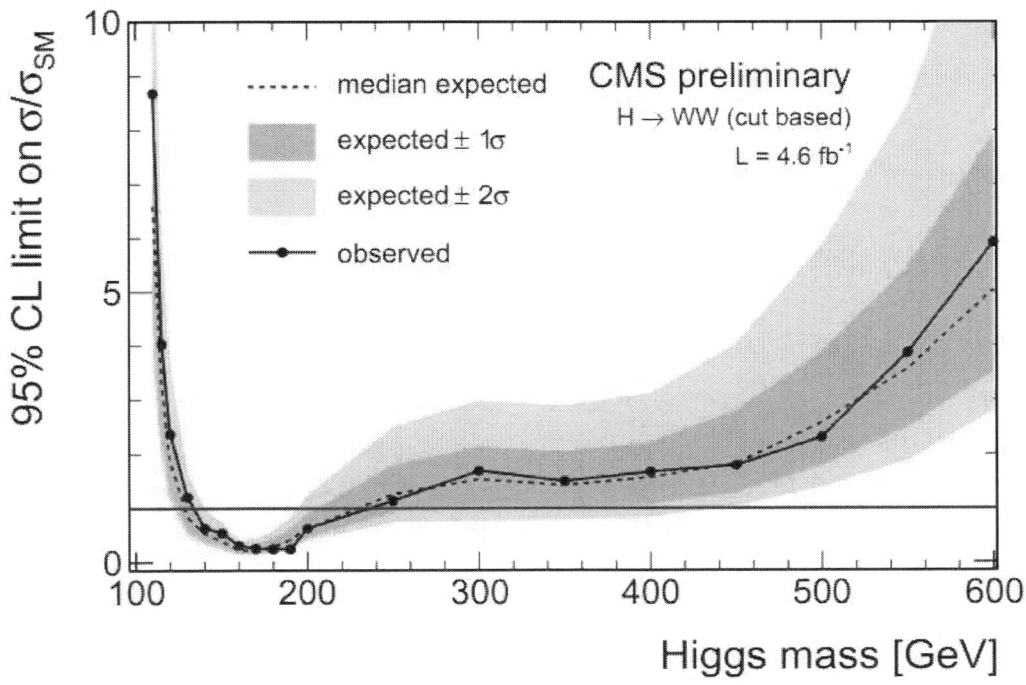

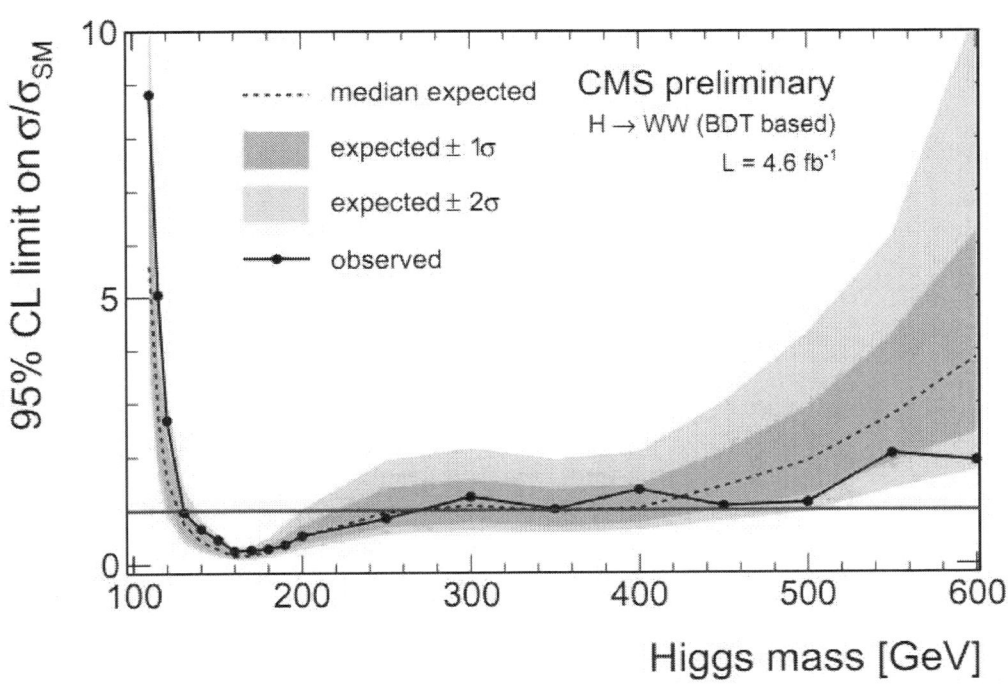

14:49

Here is the first of my unofficial combinations as the discussion time ends. This is the diphoton channels combined for ATLAS+CMS. Remember that this is approximate and you should not try to read the number of sigmas from this. I may revise it later when better version of the plots become available.

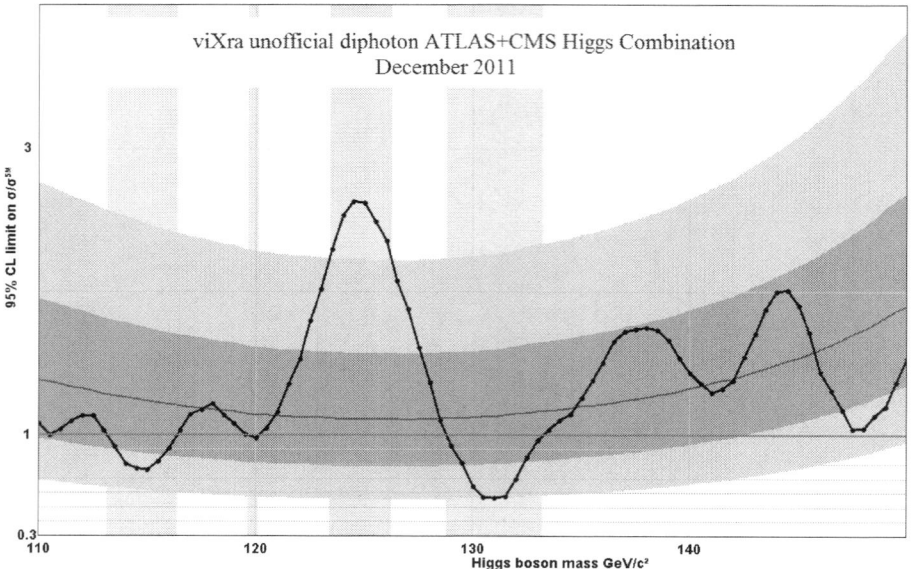

14:56

ATLAS have now released 3 new conference notes so I will update the pixtures

17:00

I have now digitised the CMS combined plot and produced this signal plot. It gives a clean indication for no Higgs about 130 GeV and the right size signal for a Higgs at about 125 GeV, but there is still noise at lower mass so chance that it could be moved.

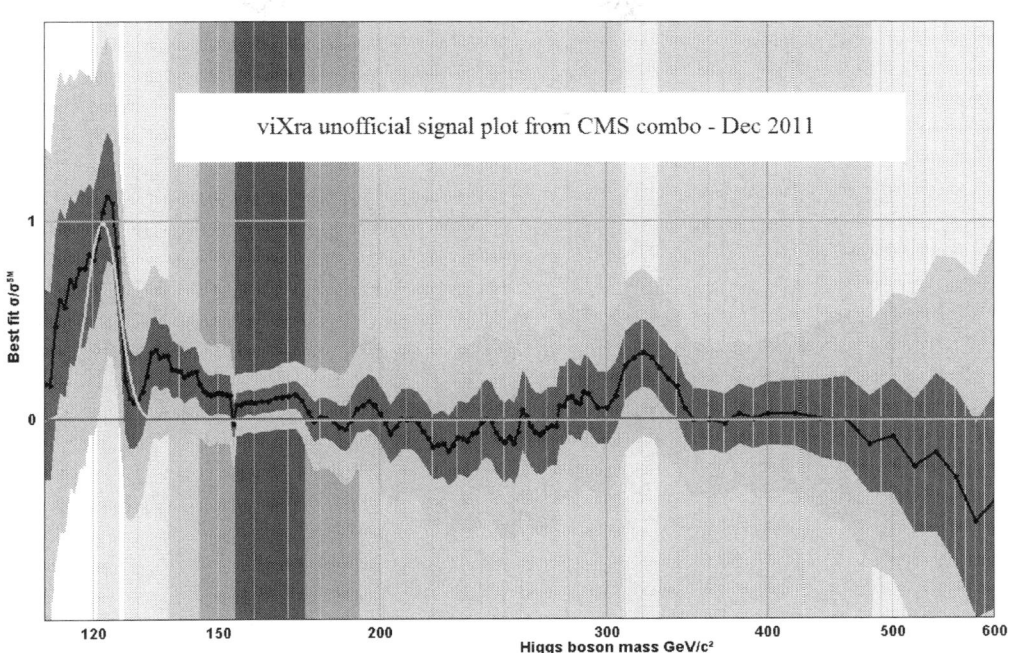

17:42

Here is the same thing for the ATLAS data

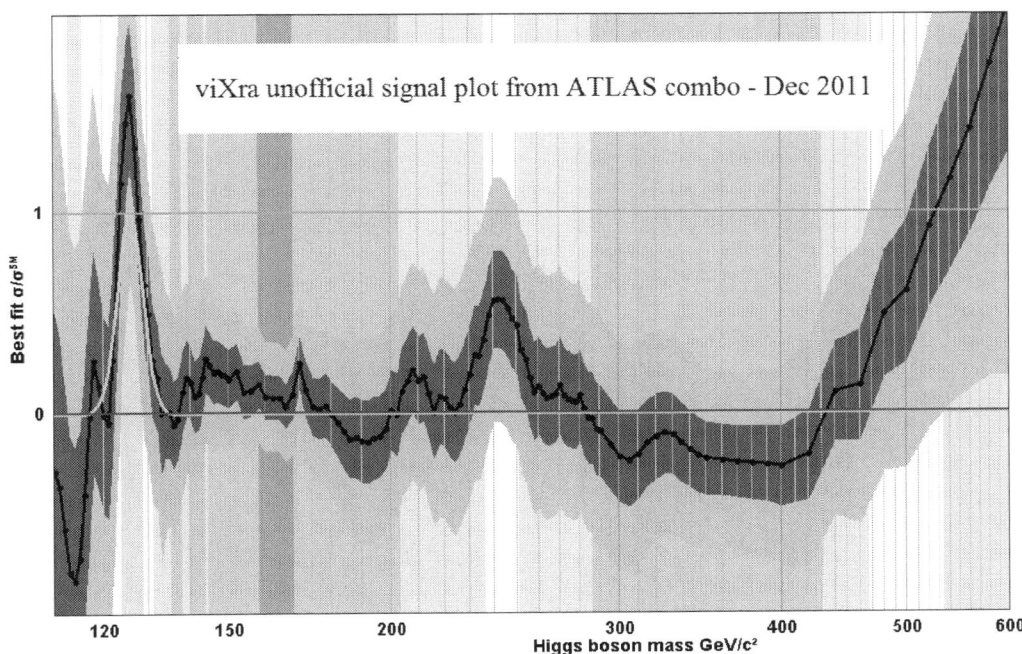

17:49

Here is the fully combined exclusion plot. The signal fits best at 124 GeV and just makes 3-sigma. Remember the official version is likely to be a little different. This is just a quick approximation.

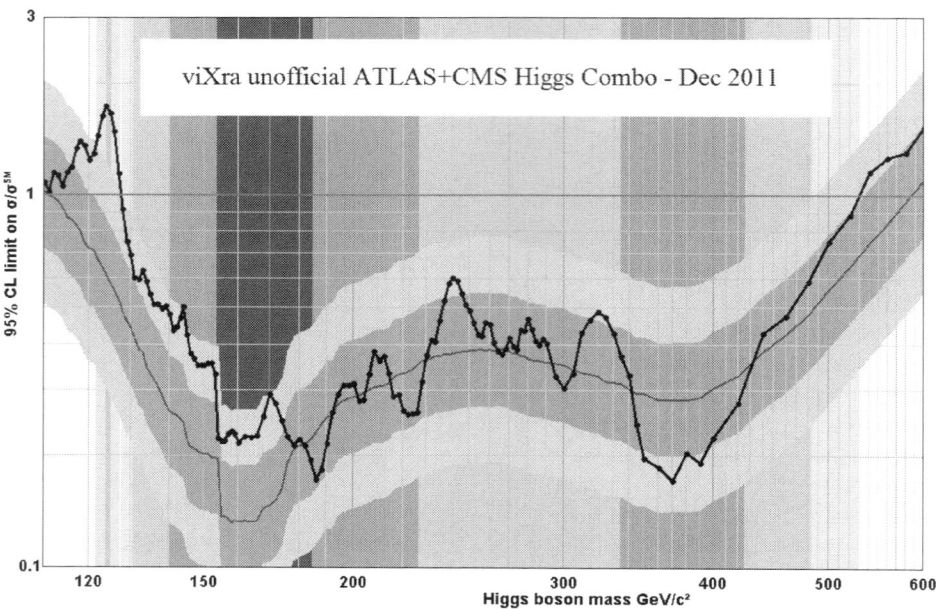

17:57

Here is the fully combined signal plot. It looks very convincing but the region below 120 GeV is not resolved yet. Until it is there will be a little room for doubt.

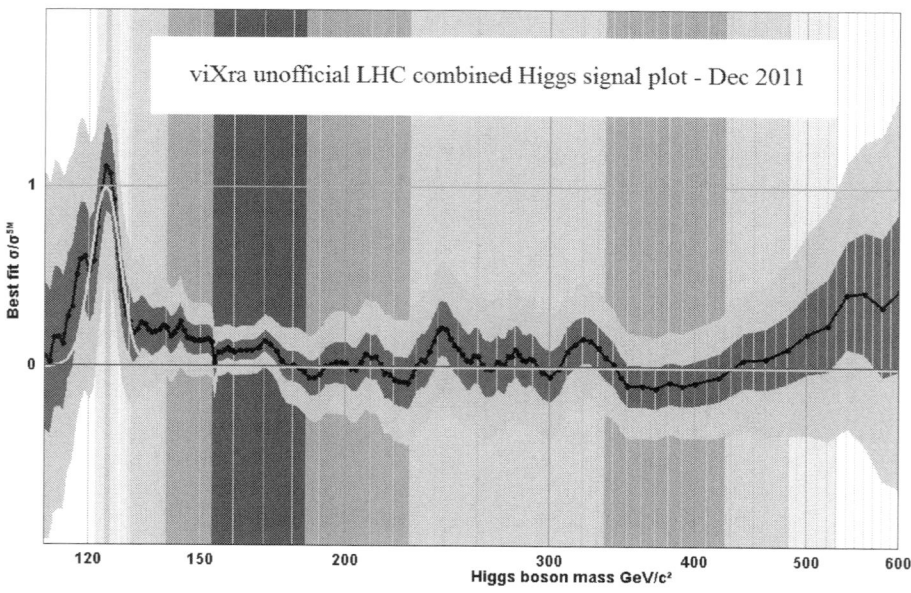

18:11

But of course we can clean up the lower region by including LEP and Tevatron too. An official combination with Tevatron data included is also planned

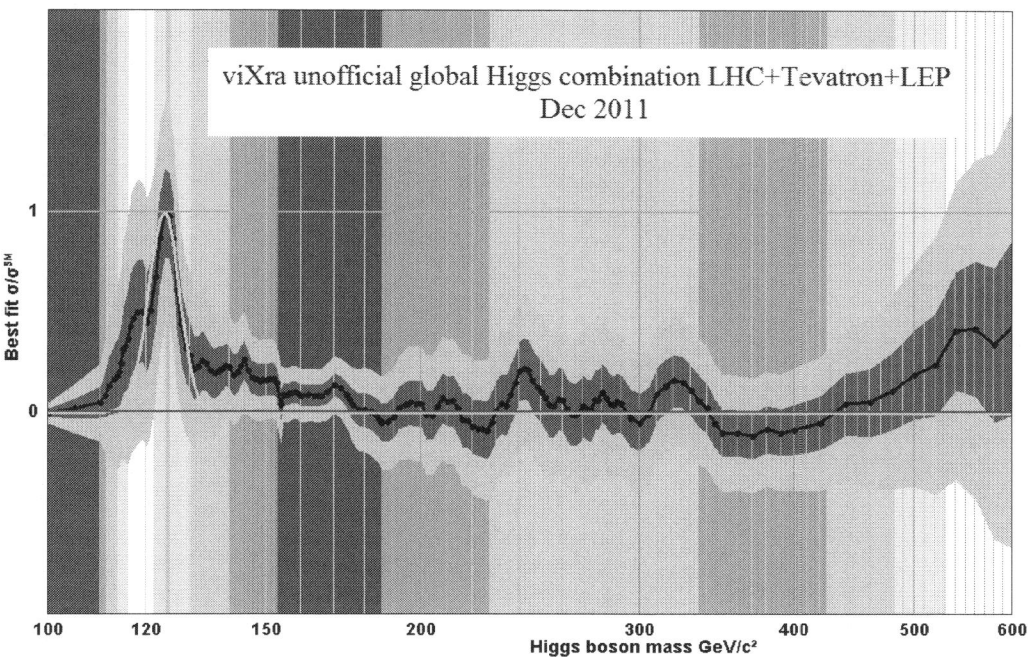

A zoomed version

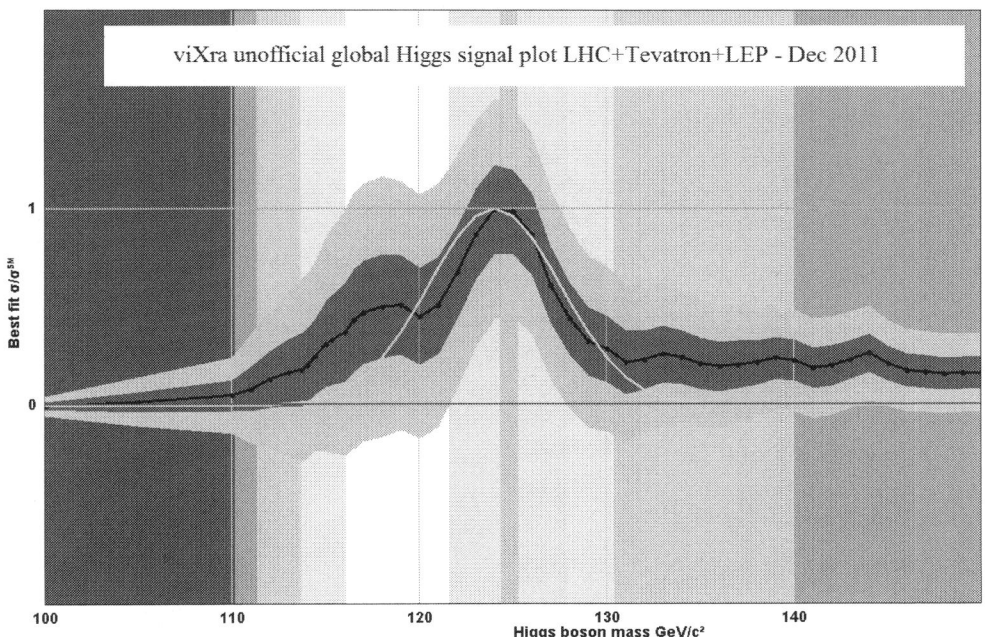

20:57

Finally here is one last combination for diphoton + ZZ in CMS and ATLAS. These are the high-resolution channels so they give a cleaner signal, but without WW the significance is less.

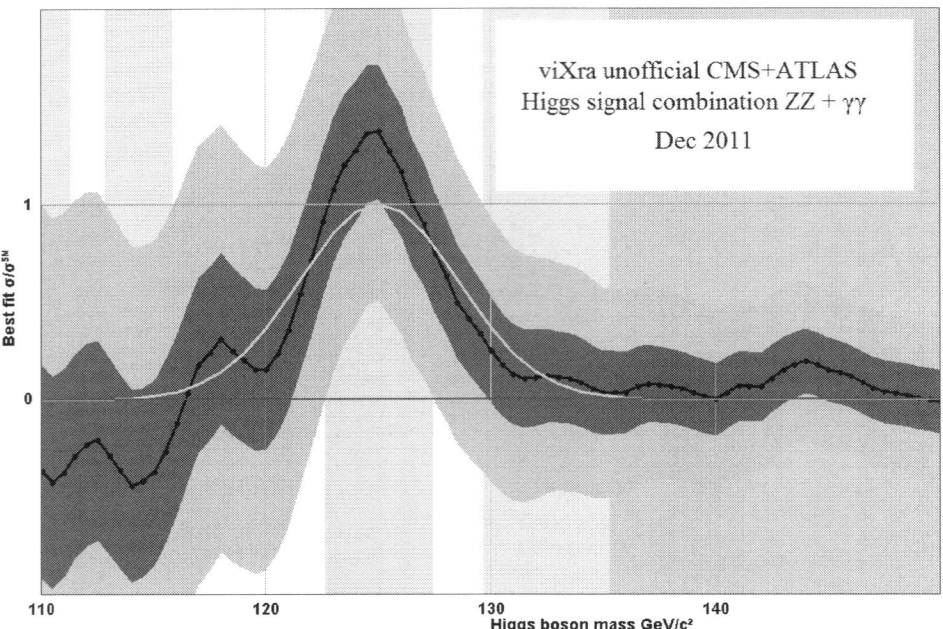

Conclusions

The result is very convincing if you start from the assumption that there should be a Higgs Boson somewhere in the range. Everywhere is ruled out except 115 GeV to 130 GeV and within that window there is a signal with the right strength at around 125 GeV with 3 sigma significance. They will have to wait for that to reach 5 sigma to claim discovery and next years data should be enough to get there or almost. I calculate that they will need 25/fb per experiment at 7 TeV to make the discovery. A big congratulations to everyone from the LHC, ATLAS and CMS who found the clear hints of Higgs when it hid in the hardest place.

I was lucky enough to meet Peter Higgs many years ago when I was a postdoc at Edinburgh and I have a big smile knowing that this has been achieved in his lifetime. Congratulations to him and the other physicists involved in discovering the mechanism of symmetry breaking. Finally, in case they are forgotten, well done also to all the phenomenologists who did the calculations to work out how the Higgs Boson could be found, not least John Ellis.

From here there is much more work to do in order to check that this particle seen today has exactly the characteristics of the Higgs, if indeed it is confirmed with more data. That will take many more years of runs at the LHC. It will also be exciting to see how this mass affects our understanding of what other physics could be in reach. I hope there are some Campaign corks popping at CERN this evening. They have had a remarkable year.

References

1. http://blog.vixra.org/2011/12/13/the-higgs-boson-live-from-cern/

Has CERN Found the God Particle? A Calculation

Philip E. Gibbs[*]

Abstract

Following the CERN announcement on December 13, 2011, physicists have been giving some very different assessments of the chances that the ATLAS and CMS detectors have seen the Higgs boson. Combining the three things I will consider, I get an overall probability for such a strong signal if there is no Higgs to be about 1 in 30. Perhaps I have failed to account for combinations where more than one of these effects could combine. That requires further coincidences but lets just call the overall result 1 in twenty. In other words, everything considered I take the observed result to be a two sigma effect.

Key Words: Higgs Boson, God Particle, CERN, LHC, ATLAS, CMS.

December 16, 2011: Has CERN Found the God Particle? A calculation

Yes I know that physicists don't use the term "God particle" but it has entered into popular culture and when the terms "Higgs Boson" and "God Particle" were trending on Twitter and Google earlier this week it was the latter that went the highest. Contrary to what some scientists imagine of the interested public, very few think that there is some religious significance attached to the particle because of this name, it's just a catchy moniker and we need not be afraid to use it.

Following the CERN announcement earlier this week, physicists have been giving some very different assessments of the chances that the ATLAS and CMS detectors have seen the Higgs boson. The CERN DG says merely that they have seen some "interesting fluctuations", while Tommaso Dorigo, (an expert on the statistical aspects of the CMS analysis) calls it "firm evidence". Theorist Lubos Motl is even more positive. He says that it is a "sure thing", but another theorist Matt Strassler has criticised such positive reports. He regards the situation as 50/50 and backed this up with a poll of experimenters that came up 9 to 1 in favour of uncertainty. This contrasts with a similar poll by Bill Murray who is lead Higgs analyst for the ATLAS collaboration. In an interview he reported a 10 to 0 vote that the Higgs had indeed been found.

What is the question?

So can we make a more objective and quantitative assessment of the current level of uncertainty over the result? You might want to know the probability that the Higgs Boson has been seen for example. Unfortunately this quantity depends on the prior probability that the Higgs Boson exists. Theoretical physicists have a very wide range of opinions on this depending on which theories they favour. Experimenters are supposed to make

---

[*] Correspondence: Philip E. Gibbs, Ph.D., Independent Researcher, UK. E-Mail: phil@royalgenes.com
Note: This report is adopted from http://blog.vixra.org/2011/12/16/has-cern-found-the-god-particle-a-calculation/

their assessments independently of such prejudices. So how can we measure the situation objectively?

Luckily there is a different question that is model independent. We can ask for the probability that the experiments would produce results as strong or stronger than those reported if there were no Higgs Boson. This conditional probability removes the theory dependence in the question so the answer should be a number that everyone could in principle agree on. The smaller this probability is, the better the certainty that the Higgs Boson has been found.

Before we can calculate the result we must define precisely what we mean by the "strength of the result". This has to be a single number so it should come from the combined results of both experiments. I will define it to be the maximum value of the CLs likelihood ratio anywhere on the plot. This takes into account both the exclusion side and the signal side of the statistics and is standard use for Higgs searches. Don't worry if you are not familiar with this quantity, it will become clearer in a minute.

Can we trust the combination?

The Higgs combination groups have tried to spread propaganda that my unofficial combinations cannot be trusted because only people familiar with the inner details of the experimental analysis are capable of doing it correctly. This is not true. I repeatedly acknowledge that my method is an approximation and that only the official combination can be used to claim a discovery, but it is a good approximation and is perfectly acceptable for making a rough assessment of the combined certainty.

They warn that people should not add the event histograms from separate experiments but that is not how my combination is done. They say that only the experts can understand the systematic uncertainties of the detectors well enough to do the combination, but these uncertainties are all built into the individual exclusion plots that they have shown and are therefore taken into account when I combine them. They warned in the past that there are correlations between the background calculations because both experiments use the same algorithms. These correlations are there and must be accounted for to get the most accurate combination possible, but they have been shown to be small. You can ignore these correlations and still get a very good approximation.

In fact the largest source of error comes from the fact that the approximate combination method assumes a flat normal probability distribution at each mass point, when in reality a more complex function based on Poisson distributions would be correct. Happily the central limit theorem says that any error function with a finite variance becomes approximately normal given high enough statistics, so the approximation gets better as more data is added.

When the combination group published their first result in November I was able to compare it with my unofficial combination done in September. This confirmed that the approximation was good. This was no surprise to me because it had already been demonstrated with the Tevatron combinations and some earlier unpublished LHC combinations. I acknowledge that my combinations for some of the individual channels were not so good because the number of events has been low, especially for the ZZ channels. This will have improved for the latest results because there is now much more data but still these individual channel combinations should be considered less certain than the overall combination.

The assessment I am doing today depends mainly on that, so this is not a big issue, however it is worth showing one further comparison between my combination and the official one for a signal channel. the plot below shows the official combination for the diphoton channel published in November when ATLAS used just 1.1/fb and CMS used just 1.7/fb. The red line is the unofficial result from viXra. It will be interesting to see how much this has improved for 5/fb.

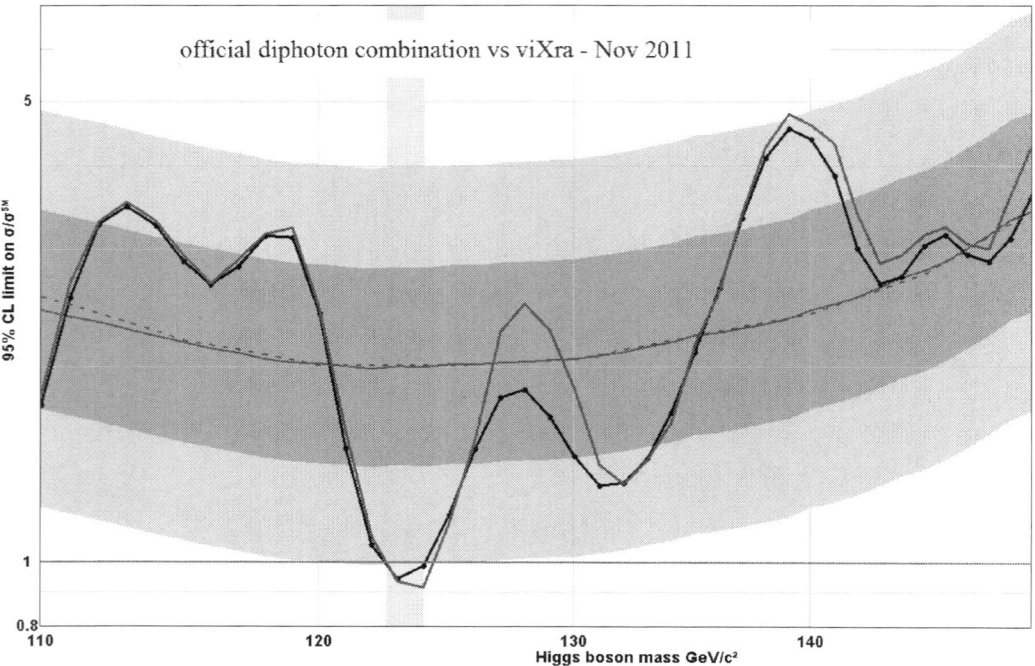

What must be evaluated?

It is possible to do a systematic evaluation of the probability in question using the combined plot. This takes into account the statistical uncertainties as well as the theoretical uncertainties in the background due to imprecise measurements of the standard model parameters (e.g. W mass) and the approximation methods used in the theory. It also includes the uncertainty in the energy resolution and other similar uncertainties in detector performance. All these things have been considered by the experts from the experimental collaborations and built into the plots, so we don't need to know the details to do the calculation (If anyone tries to claim otherwise they are wrong)

However, there is also the possibility that the experimenters have made some more fundamental kind of error. There may be a subtle fault in the detectors that has not shown up in all the calibration tests which causes an excess on the plot where there should not be one. This should not happen because there are hundreds of people checking for such errors and they are all very competent. Nevertheless bad luck can strike and throw everything out. This has been the case before and it is probably the case with the OPERA result indicating that neutrinos are faster than light.

A second similar possibility is that the theorists have underestimated the accuracy of some of their calculations so that the background calculation is a little off in one mass range. The analysis involves subtracting a very small signal from a large background, especially in the diphoton channel, so the scope for magnifying any inaccuracy has to be considered. A miscalculation of the signal size is also possible but less likely to lead to a bad result.

As I said, the published plots include all the known experimental and theoretical uncertainties, but these other unknown errors in experiment and theory cannot be accounted for exactly. They can only be estimated based on past experience. Some "expert theorists" say that us more "naive theorists" don't appreciate these facts. Do we really sound so stupid?

What is the chance of an experimental fault?

How often do experimental faults contribute to a false positive like the excess reported this week? We can only look at past performance but I am not aware of any careful surveys, so a guestimate is required. Someone else may be able to do better. The answer might be one in a hundred but let's be more conservative and say one in ten. If you think it is more common please fell free to reevaluate for yourselves.

However, with the CERN Higgs result we have good evidence that such a fault is not the cause of the excess. That is because there are two independent experiments reporting a very similar result. ATLAS and CMS may seem very similar from the answers they produce, but the detector technologies they use are quite different. The chance of a common fault producing the excess in both detectors must therefore be very small. I am going to assume that this is negligible. If anyone thinks otherwise please explain why.

This means that if the excess is due to such a fault it must be a coincidence that it has a similar effect for both experiments. If there is a one in ten chance of a fault for one experiment, the chance for two independent experiments is one in 100, but even then that is the chance that they would produce the fault at different places. Lets have a look at the two signal plots together.

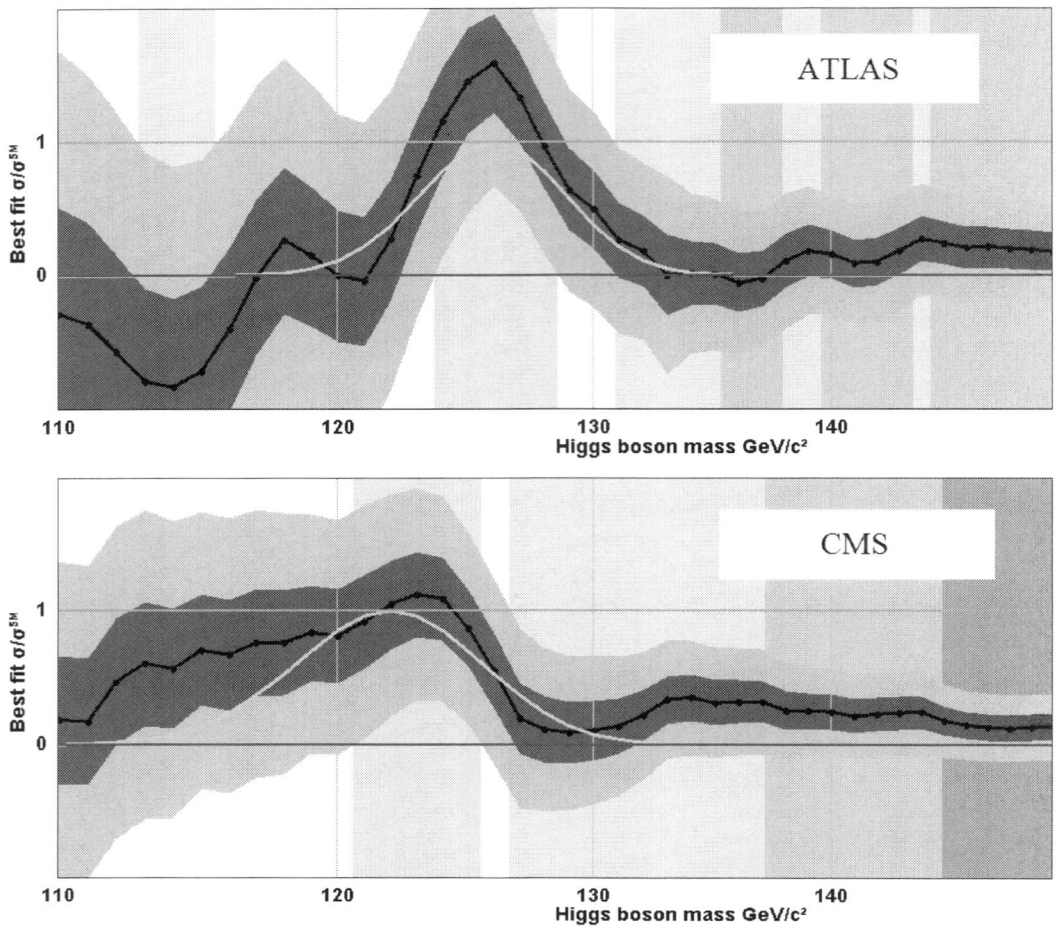

The positions on the maximum excess differ by about 2 GeV but the mass resolution is around 2% so this is not an inconsistency. If these excesses are produced by detector faults then the chance of them lining up so close would be small. How small? That depends on some unknowns. we can't just say the fault could appear anywhere in the mass range, so let's be conservative and just call it a one in three chance.

Overall then we arrive at a one in 300 chance for the observed excess to be explained by a coincidental combination of detector faults. I think this is conservative. Someone else might estimate it to be more probable.

What is the chance of a theoretical Fault?

The other outside possibility is that the result has been afflicted by a misunderstood background so that the observed excess is really just a subtle effect of the Higgsless standard model that the theorists failed to recognize or estimate correctly. Again this is unlikely but it happens and must be considered. How often does it happen? Once in a hundred perhaps? I will be more cautious and assume one in ten. You may think that is an underestimate in which case you can make your own evaluation.

But again we have more than one place to look. The separate experiments could well be affected by the same theoretical error but the different decay channels are much more independent. There may be some small chance that a single theoretical error could affect all the channels but this would have a small probability, say one in a hundred. If you think it is bigger please justify how that could happen.

So now let's look at the combined signal plots for the three main channels; diphoton, ZZ->4l and WW->lvlv. For the WW plot I can't use the latest CMS results because the plots shown are frankly rubbish quality. I hope they will improve them before publication. However the WW channel has good sensitivity even with less data so I will show the combination from the summer.

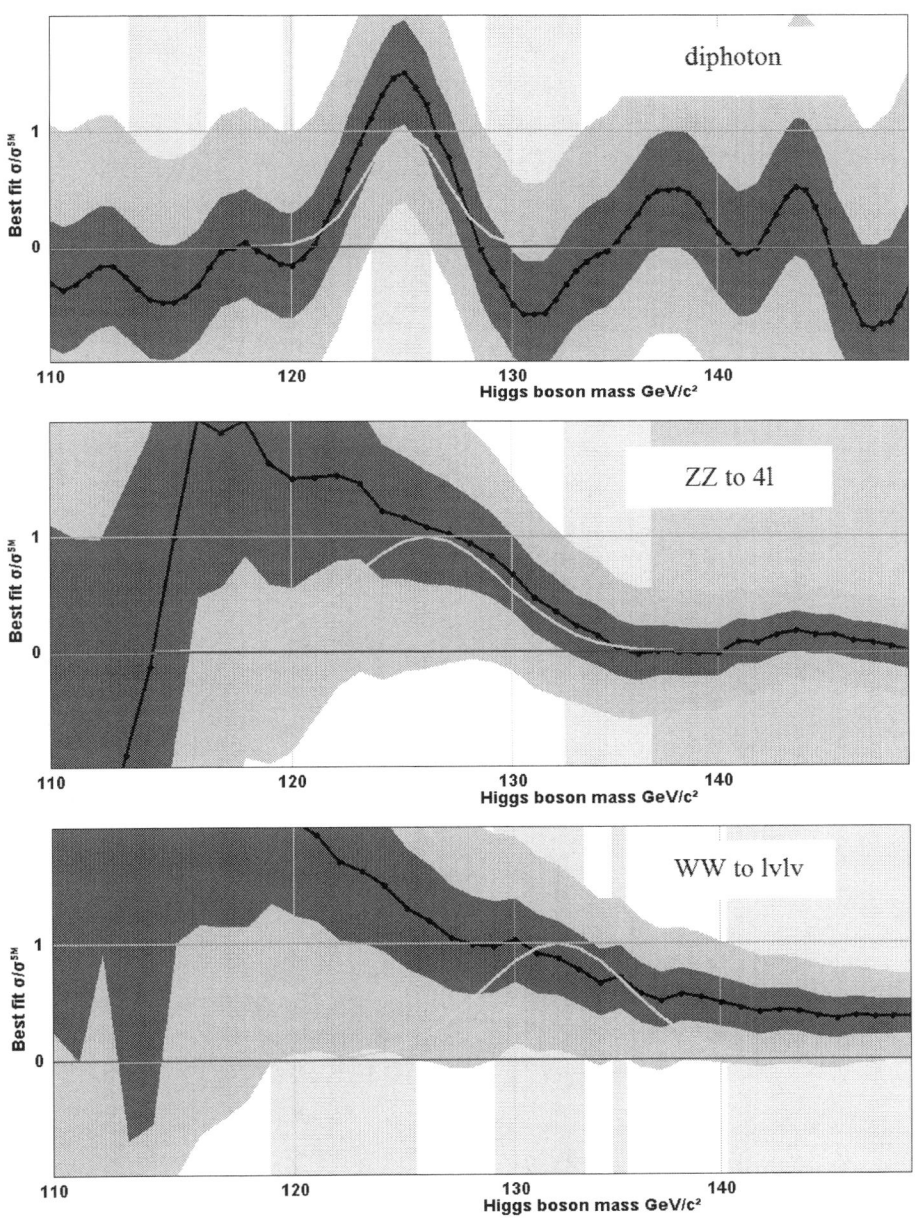

All three channels show an excess in the same low mass region so if this is due to independent faults it would require a coincidence. However, the excess is not as good in ZZ and WW as in the diphoton channel. I am going to put the probability at one in a hundred overall and add to this the probability of one in a hundred for a common fault that affects all three. So the overall chance for a fault from theory is one in 50. Some people will say that this is a low estimate and some people will say that it is low. Others will say that it is nonsense to attempt such an estimate. Never mind, I am just giving it my best honest shot. Let others do the same.

What is the chance of a statistical fluctuation?

The last thing to consider is what is the probability of getting s signal as string or stringer than that observed according to the statistical analysis. Actually this also takes into account some theoretical uncertainty and measurement error, but mostly it is statistical. This is a probability that can be worked out more scientifically, but it does include the Look Elsewhere Effect which is partly subjective.

First consider what would be the chance of seeing a signal as strong as the one reported at the fixed mass point of the maximum excess if in fact there was no Higgs Boson. The plot shows a three sigma excess at 124-125 GeV. This would have been much stronger if the peaks from the two experiments had coincided more closely, possibly about 4 sigma. This discrepancy may be due to some detector calibration that could be corrected but it is correct that we do not take that possibility into account. The 3 sigma excess is what we should work with.

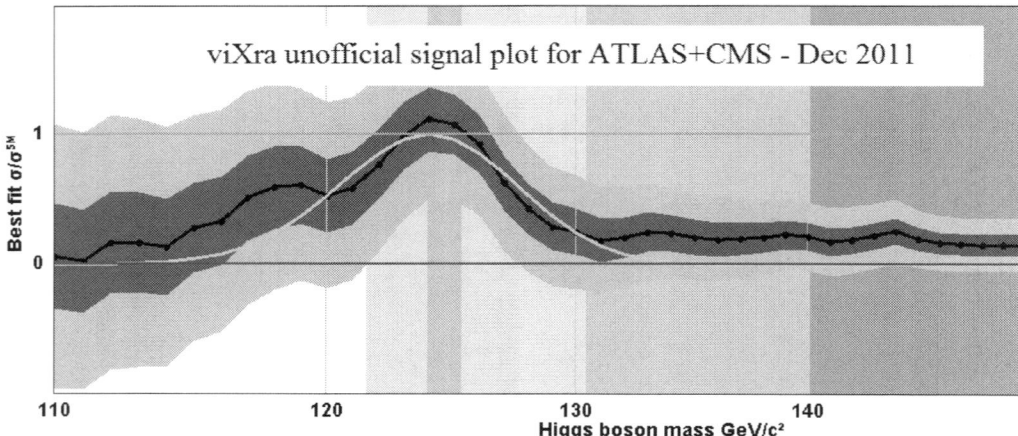

As everyone knows, the probability of a three sigma fluctuation is one in 370, but that allows for fluctuations up or down. So the probability for an excess this big or stronger at this point is one in 740. But we need to know the probability for an excess this strong anywhere on the plot. In other words we need to multiply by the Look Elsewhere Effect factor. Have a look at the plot over the entire range

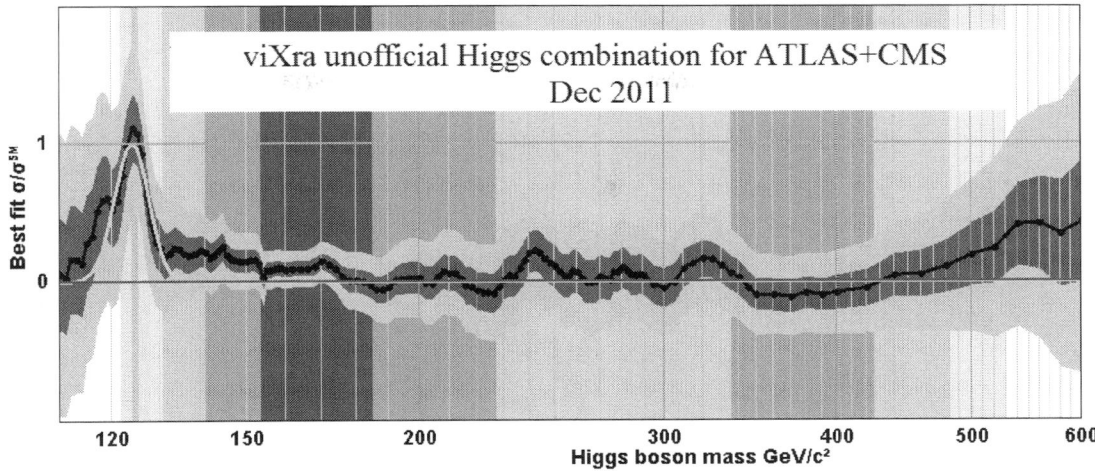

Notice that for the entire range from 130 GeV to 600 GeV the line remains within 2 sigma of the zero line. Big deviations are indeed rare but how rare?

Another point to consider is that if there was a three sigma fluctuation at say 180GeV, the Higgs would still be excluded at that point. This would not count as such a strong signal. This is why I specified that the strength should be measured using the CLs statistic which takes the ration of the probability for the signal hypothesis over the probability from the null hypothesis. This means that the probability of getting a signal as strong in the regions where the Higgs is excluded is much smaller. In fact we can neglect this altogether. So we need only count the regions from 114 GeV (using LEP) to about 135 GeV and perhaps 500 to 600 GeV. Hoe big is the LEE factor for these regions. This depends on the width of the signal which we see to be about 5 GeV in the low mass range due to mass resolution of the detector, and which is much bigger above 500 GeV due to a very large natural width for a high mass Higgs Boson. The LEE factor will therefore be about 6 but let's call it 10 to be extra cautious.

This gives a final answer for the probability of a fluctuation to be about one in 70.

The final answer?

Combining the three things I have considered I get an overall probability for such a strong signal if there is no Higgs to be about 1 in 30. Perhaps I have failed to account for combinations where more than one of these effects could combine. That requires further coincidences but lets just call the overall result 1 in twenty. In other words, everything considered I take the observed result to be a two sigma effect.

What about prior probabilities?

there is one more thing you need to take into account when considering how likely a result of any number of sigmas significance is going to stand the test of time. That is your prior estimate for the probability of it being true. The OPERA neutrino observation is a good example of an extreme case. A six sigma effect was observed, but he prior probability of neutrinos going faster than light would be considered very small by most theoretical physicists. It follows that the probability for this result to go away is quite high despite the statistical significance. An experimental fault is likely to be the biggest contributing factor despite the care of the experimenters.

In fact most 3 sigma excesses for observations beyond the standard model do go away. This is because the prior chance of any one such effect being correct is very small. You can consider this to be part of the Look Elsewhere Effect too. However, the observation of the Higgs Boson is a very different case. Most theoretical physicists would estimate the prior probability for the existence of a Higgs (like) Boson is very high. The standard model provides a very simple explanation of electroweak symmetry breaking but there is no simple way to understand a Higgsless universe. This makes the prior probability high which means that the chance of the 2 sigma result going away is small. There is a bigger chance however that it could move to a different mass.

Not everyone agrees with this. Some people do not think that the Higgs Boson can exist. Stephen hawking is one of them. These people would assign a low value to the prior probability that the signal for the Higgs will be seen and so they will consider it very likely that the present observation will go away. I doubt that there are enough people of this opinion to account for much doubt among the experimenters.

How long will it take to settle this?

To claim a discovery the combined results must give a 5 sigma excess without considering the Look Elsewhere Effect. How long this takes depends on a certain amount of luck. If the peaks of the excesses comes closer together with more data, then the excess will grow faster than you would otherwise expect. In that case the matter might be settled with just twice as much data and the whole thing will be over by the summer. On the other hand, if they are unlucky it could easily require the full dataset from 2012 to get enough data to finish the job properly. It will then not be until March 2013 when the combination is ready that they will finally be able to declare a discovery.

References

1. http://blog.vixra.org/2011/12/16/has-cern-found-the-god-particle-a-calculation/

Article

Electron Spin Precession for the Time Fractional Pauli Equation

Hosein Nasrolahpour[*]

Dept. of Physics, Faculty of Basic Sci., Univ. of Mazandaran, P.O. Box 47416-95447, Babolsar, Iran
Hadaf Institute of Higher Education, P.O. Box 48175 – 1666, Sari, Mazandaran, Iran

Abstract

In this work, we aim to extend the application of the fractional calculus in the realm of quantum mechanics. We present a time fractional Pauli equation containing Caputo fractional derivative. By use of the new equation we study the electron spin precession problem in a homogeneous constant magnetic field.

Keywords: fractional quantum mechanics, time fractional Pauli equation, electron spin precession.

1. Introduction

Fractional derivatives provide a relevant tool for understanding the dynamics of the complex systems. Recently fractional calculus has found interest and applications in the domain of quantum mechanics and field theory (see [5] and refs. in). Fractional quantum mechanics (FQM) is in fact generalization of standard quantum mechanics to discuss quantum phenomena in fractal and complex systems. Up to now, several different approaches on FQM have been investigated [6-19].

In FQM, space fractional Schrödinger equation has been obtained by Laskin [9]. He constructed the fractional quantum mechanics using the Lévy path integral and considered some aspects of the space fractional quantum system. Space fractional Schrödinger equation similar to the standard one satisfies the Markovian evolution law. However, one can introduce the time fractional Schrödinger equation to describe non- Markovian evolution in quantum mechanics [10].

In Refs. [10, 11] some properties of time fractional Schrödinger equation have been studied. It is showed that Hamiltonian in this case is non-Hermitian and non-local in time. Furthermore, it is found [10] that for free particle case the wave function evolves in such a way that the value of total probability increases to the value $\frac{1}{\alpha^2}$ (where α is the order of derivative in the time fractional Schrödinger equation) in the limit $t \to \infty$. The probability structure of time fractional Schrödinger equation in the case of low-level fractionality [20, 21] also has been discussed in [11]. It has been proved that for the basis state the probability density increases initially, and then fluctuates around a constant value in later times.

In the standard quantum mechanics, the well-known Pauli equation (the non-relativistic evolution equation of spin $\frac{1}{2}$ particles) is given by [1]

---

[*] Correspondence: Hosein Nasrolahpour, Department of Physics, Faculty of Basic Sciences, University of Mazandaran, P. O. Box 47416-95447, Babolsar, Iran. E-mail: hnasrolahpour@gmail.com

$$[\frac{1}{2m}(\hat{P}-\frac{e}{c}A)^2 + e\phi + \mu_B \hat{\sigma}.B]\Psi = i\hbar\frac{\partial\Psi}{\partial t} \tag{1}$$

Where A is the three-component magnetic vector potential and ϕ is the electric scalar potential and $\Psi = \begin{pmatrix}\Psi_1\\\Psi_2\end{pmatrix}$ is the two component spinor wave function and $\mu_B = \frac{|e|\hbar}{2mc}$ is the so-called Bohr magneton.

As we mentioned above, quantum phenomena in disordered systems such as random fractal structure can be discussed by the fractional quantum mechanics. In disordered systems spatial and temporal fluctuations often obey Gaussian statistics due to the central limit theorem. However, it has been found in recent years that anomalous broad Lévy type distributions of the fluctuations can occur in many complex systems [24].With this motivation we aim to present a fractional Pauli equation containing Caputo fractional derivative and study the electron spin precession within this framework.

In the following, fractional calculus [2-4] is briefly reviewed in Sec. 2. The time fractional Pauli equation is presented in Sec. 3. The time-dependent spin function and expectation values for the spin observables are obtained in terms of the Mittag-Leffler functions. The probability for spin-up and spin-down also are calculated in this section. In Sec. 4, we will present our conclusions.

2. Fractional calculus

2.1. The Caputo fractional derivative operator

Fractional calculus (FC) has a long history, when the derivative of order $\alpha = \frac{1}{2}$ has been described by Leibniz in a letter to L'Hospital in 1695 [3]. FC is the calculus of derivatives and integrals with arbitrary order, which unify and generalize the notions of integer order differentiation and n-fold integration, which have found many applications in recent studies to model a variety processes from classical to quantum physics [4-21]. In contrast with the definition of the integer order derivative, the definition of the fractional one is not unique. In fact, there exists several definitions including, Grünwald - Letnikov, Riemann - Liouville, Weyl, Riesz and Caputo for fractional order derivative. Fractional differential equations defined in terms of Caputo derivatives require standard boundary (initial) conditions. For this reason, in this paper we only use the Caputo fractional derivative. The left (forward) Caputo fractional derivative of a function f (t) is defined by

$${}_0^c D_t^\alpha f(t) = \frac{1}{\Gamma(n-\alpha)}\int_0^t (t-\tau)^{n-\alpha-1} f^{(n)}(\tau)d\tau, \quad \alpha > 0, t > 0 \tag{2}$$

Where n is an integer number and α is the order of the derivative such that $n-1 < \alpha < n$ and $f^{(n)}(\tau)$ denotes the n-th derivative of the function $f(\tau)$. Laplace transform to Caputo's fractional derivative gives

$$L\{{}_0^c D_t^\alpha f(t)\} = s^\alpha F(s) - \sum_{m=0}^{n-1} s^{\alpha-m-1} f^{(m)}(0) \tag{3}$$

Where $F(s)$ is the Laplace transform of $f(t)$.

2.2. The Mittag-Leffler function

The Mittag-Leffler function is a generalization of the exponential function, which plays an important role in the fractional calculus. The Mittag-Leffler function is such a one-parameter function defined by the series expansion as

$$E_\alpha(z) = \sum_{k=0}^{\infty} \frac{z^k}{\Gamma(1+\alpha k)} \qquad \alpha \in \mathbb{C}, \alpha > 0, z \in \mathbb{C} \tag{4}$$

And its general two-parameter representations is defined as

$$E_{\alpha,\beta}(z) = \sum_{k=0}^{\infty} \frac{z^k}{\Gamma(\beta+\alpha k)} \qquad \alpha,\beta \in \mathbb{C}, \alpha,\beta > 0, z \in \mathbb{C} \tag{5}$$

Where $\mathbb{C}$ is the set of complex numbers and $\Gamma(\alpha)$ denotes the gamma function. Laplace transform for Mittag-Leffler function is very useful in solving fractional differential equations. The Laplace transformations for several Mittag-Leffler functions are summarized below

$$L\{E_\alpha(-\lambda t^\alpha)\} = \frac{s^{\alpha-1}}{s^\alpha + \lambda} \tag{6a}$$

$$L\{t^{\alpha-1} E_{\alpha,\alpha}(-\lambda t^\alpha)\} = \frac{1}{s^\alpha + \lambda} \tag{6b}$$

$$L\{t^{\beta-1} E_{\alpha,\beta}(-\lambda t^\alpha)\} = \frac{s^{\alpha-\beta}}{s^\alpha + \lambda} \tag{6c}$$

Where $s > |\lambda|^{\frac{1}{\alpha}}$.

3. Time fractional Pauli equation

As we mentioned in section 1, one can introduce the time fractional Schrödinger equation to describe non-Markovian evolution [*] in quantum mechanics. In this section we generalize the time fractional Schrödinger equation [12, 13] and obtain the following time fractional Pauli equation

$$[\frac{1}{2m}(\hat{P} - \frac{e}{c}A)^2 + e\phi + \mu_B \hat{\sigma}.B]\Psi = i\hbar_\alpha \frac{\partial^\alpha \Psi}{\partial t^\alpha} \qquad 0 < \alpha \leq 1 \tag{7}$$

Where $\hbar_\alpha = M_P c^2 T_P^\alpha$ is a scaled Planck constant. The parameters M_P and T_P are Planck mass and Planck time, respectively, which are defined as

---

[*] Non-Markovian systems appear in many branches of physics, such as quantum optics, solid state physics, quantum information processing, and quantum chemistry (see for example the Ref. [22]).

$$T_P = \sqrt{\frac{G\hbar}{c^5}} \quad , \quad M_P = \sqrt{\frac{c\hbar}{G}} \tag{8}$$

Where G and c are the gravitational constant and the speed of light in vacuum, respectively.

We now use the Eq. (7) to discuss the electron spin precession problem in a homogeneous constant magnetic field. For this purpose, we first assume that, the electron is fixed at a certain location and its spin is the only degree of freedom. Also, let the magnetic field consist of a constant field $\vec{B}$ in the Z direction

$$\vec{B} = B_0 \hat{k}$$

Therefore, that part of the time fractional Pauli equation Eq. (7), which contains the spin yields

$$(i\hbar_\alpha)^c_0 D^\alpha_t \chi = \hbar_\alpha \omega^\alpha_L \hat{\sigma}_z \chi \tag{9}$$

Where $\omega_L = -\frac{eB}{2mc}$ are the so-called Larmor frequency and the Pauli matrices for a spin $\frac{1}{2}$ particles are as below

$$\hat{\sigma}_y = \begin{pmatrix} 0 & -i \\ i & 0 \end{pmatrix} \quad , \quad \hat{\sigma}_z = \begin{pmatrix} 1 & 0 \\ 0 & -1 \end{pmatrix} \quad \hat{\sigma}_x = \begin{pmatrix} 0 & 1 \\ 1 & 0 \end{pmatrix}$$

Science the Hamiltonian of our system is a 2x2 matrix, the spin function in arbitrary time t must be written as a column matrix of two components and can be derived as below,

$$\chi_\alpha(t) = \begin{pmatrix} a(t) \\ b(t) \end{pmatrix} = \begin{pmatrix} e^{i\gamma} \cos\left(\frac{\theta}{2}\right) E_\alpha(-i(\omega_L t)^\alpha) \\ e^{i\delta} \sin\left(\frac{\theta}{2}\right) E_\alpha(i(\omega_L t)^\alpha) \end{pmatrix} \tag{10}$$

Where γ and δ are arbitrary phase constants.

Now, we able to calculate the expectation values for the observables $\hat{s}_x, \hat{s}_y, \hat{s}_z$. Then we will have

$$<\hat{s}_x>_{\alpha,t} = \frac{\hbar}{2}(\chi^\dagger \hat{\sigma}_x \chi) = \frac{\hbar}{4}\sin(\theta)[e^{i(\delta-\gamma)}(E_\alpha(i(\omega_L t)^\alpha))^2 + e^{-i(\delta-\gamma)}(E_\alpha(-i(\omega_L t)^\alpha))^2] \tag{11}$$

$$<\hat{s}_y>_{\alpha,t} = \frac{\hbar}{2}(\chi^\dagger \hat{\sigma}_y \chi) = \frac{i\hbar}{4}\sin(\theta)[e^{-i(\delta-\gamma)}(E_\alpha(-i(\omega_L t)^\alpha))^2 - e^{i(\delta-\gamma)}(E_\alpha(i(\omega_L t)^\alpha))^2] \tag{12}$$

$$<\hat{s}_z>_{\alpha,t} = \frac{\hbar}{2}(\chi^\dagger \hat{\sigma}_z \chi) = \frac{\hbar}{2}\cos(\theta)[E_\alpha(-i(\omega_L t)^\alpha) E_\alpha(i(\omega_L t)^\alpha)] \tag{13}$$

Here we can see explicitly that as $\alpha \to 1$, Eq. (13) gives

$$<\hat{s}_z>_{\alpha=1,t} = \frac{\hbar}{2}\cos(\theta)[E_1(-i(\omega_L t))E_1(i(\omega_L t))] = \frac{\hbar}{2}\cos(\theta) \qquad (14)$$

So we have $<s_z>_{\alpha=1,t} = <s_z>_{\alpha=1,t=0}$, as expected from the standard quantum mechanics. Furthermore, for the special case of $\alpha = \frac{1}{2}$, we have

$$<\hat{s}_z>_{\alpha=\frac{1}{2},t} = \frac{\hbar}{2}\cos(\theta)e^{-2\omega_L t}erfc(i\sqrt{\omega_L t})[2 - erfc(i\sqrt{\omega_L t})] \qquad (15)$$

Where $erfc(z)$ denoted the complementary error function [23], which is defined by

$$erfc(z) = \frac{2}{\sqrt{\pi}}\int_z^\infty e^{-t^2}dt \qquad (16)$$

Also by use of Eq. (10) we can calculate the probability for spin-up, $P_{\alpha\uparrow}$, and spin-down, $P_{\alpha\downarrow}$, at $t > 0$.

So we have

$$P_{\alpha\uparrow} = |a(t)|^2 = \cos^2(\frac{\theta}{2})[E_\alpha(-i(\omega_L t)^\alpha)E_\alpha(i(\omega_L t)^\alpha)] \qquad (17)$$

$$P_{\alpha\downarrow} = |b(t)|^2 = \sin^2(\frac{\theta}{2})[E_\alpha(-i(\omega_L t)^\alpha)E_\alpha(i(\omega_L t)^\alpha)]. \qquad (18)$$

4. Discussion and Conclusion

Fractional quantum mechanics is a generalization of traditional quantum mechanics to discuss quantum phenomena in disordered systems such as random fractal structure. In disordered systems spatial and temporal fluctuations often obey Gaussian statistics due to the central limit theorem. However, it has been found in recent years that anomalous broad Lévy type distributions of the fluctuations can occur in many complex systems [24]. With this motivation a time fractional Pauli equation is introduced.

In this paper we have studied the spin precession in the framework of fractional dynamics. For this purpose: first, we have presented the time fractional Pauli equation. Utilizing this equation we have derived the time-dependent spin function for our system. Then using the Eq. (10), we have obtained the expectation values for the observables $\hat{s}_x, \hat{s}_y, \hat{s}_z$. The following result can be easily deduced:

By use of the standard Pauli equation we have,

$$<\hat{s}> = \frac{\hbar}{2}(\cos(2\omega_L t + \delta - \gamma)\sin(\theta), \sin(2\omega_L t + \delta - \gamma)\sin(\theta), \cos(\theta)) \qquad (19)$$

Obviously, the above relation can be deduced from Eqs. (11-13) as $\alpha \to 1$ and we will have a conserved spin component in the field direction while the spin precesses around the z axis with twice the Larmor frequency $2\omega_L$ (we know that this is due to the gyromagnetic factor 2 of the spin). But as we can see from the Eq. (13) we have:

$$(20)$$

$$<\hat{s}_z>_{\alpha,t} \neq <\hat{s}_z>_{\alpha,t=0}, \quad 0<\alpha<1$$

These results show that spin component in the field direction s_z is not conserved for the arbitrary case of α. Therefore, spin precession phenomena is complicated in the framework of fractional dynamics(also as a result, we can consider modifications to the gyromagnetic factor of the spin in this framework). Furthermore, we have calculated the time dependent probability for spin-up, $P_{\alpha\uparrow}$, and spin-down, $P_{\alpha\downarrow}$ at $t>0$, for the arbitrary case of α ($0<\alpha<1$). As a consequence of these results we can predict a transition from one spin state to the other spin state in our system, when time flows.

In this paper we only consider the case of $0<\alpha<1$, it is also interesting to consider the case of $1<\alpha<2$ for the Eq. (7) and derive new results. We hope to do this in future work.

Recently fractional calculus has found interest and applications in the context of relativistic quantum mechanics and field theory [25-32]. We hope to study some applications of fractional calculus in these interesting areas of physics, as well.

Acknowledgement: The author is grateful to Richard Herrmann for many useful discussions and explanations.

References

[1] J. J. Sakurai, Modern Quantum Mechanics *(Addison-Wesley, New York,* 1994).
   W. Greiner, Quantum Mechanics: An Introduction, 4th ed. *(Springer, Berlin,* 2001).
[2] I. Podlubny, Fractional Differential Equations *(Academic Press, New York,* 1999).
[3] K. S. Miller, B. Ross, An introduction to the fractional calculus and fractional differential equations (*John Wiley & Sons, Inc.,* 1993).
   K. B. Oldham, J. Spanier, the Fractional Calculus *(Academic Press, NewYork,* 1974).
[4] R. Hilfer, Applications of Fractional Calculus in Physics (*World Scientific,* Singapore, 2000).
   G. M. Zaslavsky, Hamiltonian Chaos and Fractional Dynamics *(Oxford Univ. Press, Oxford,* 2005).
   R. Metzler and J. Klafter, *Phys. Rep.,* **339**(2000) 1–77.
   R. Metzler and J. Klafter, *J. Phys. A,* **37**(2004)161–208.
[5] H. Nasrolahpour, Applications of fractional derivative in particle physics (MSc Thesis), Published By *University of Mazandaran,* 2008.
[6] L. Nottale , Fractal space-time and micro-physics (*World Scientific, Singapore,*1993).
   L. Nottale, *Prog. Phys.* **1** (2005) 12.
   M. N. Celerier , L. Nottale, *J. Phys. A Math. Gen.* **37** (2004) 931.
[7] N. Laskin, *Phys.Lett. A* **268** (2000) 298.
[8] N. Laskin, *Phys. Rev. E* **62** (2000) 3135.
[9] N. Laskin, *Phys. Rev. E* **66** (2002) 056108.
[10] M. Naber, *J. Math. Phys.* **45**(2004) 3339-3352.
[11] A. Tofighi, *Acta Phys. Pol. A.* Vol. **116** (2009)114-118.
[12] M. Bhatti, *Int. J. Contemp. Math. Sci.,* **2** (2007) 943.
[13] S. I. Muslih ,Om P. Agrawal , Dumitru Baleanu , *Int. J. Theor. Phys.* **49** (2010) 1746–1752.
[14] R. Herrmann, *Physica A.* **389** (2010) 693-704.
[15] R. Herrmann, *J.Phys.G: Nucl. Part. Phys,* **34** (2007) 607.
[16] R. Herrmann, *Physica A.* **389** (2010) 4613-4622.

[17] E. Goldfain, *Comm. Non. Sci. Num. Siml.* **13** (2008) 666–676.
[18] E. Goldfain, *Chaos, Solitons & Fractals*, **22**(2004) 513-520.
[19] V. E. Tarasov, *Physics Letters A* **372** (2006) 2984-2988.
   V. E. Tarasov, *Theor. and Math. Phys.*, **158**(2) (2009) 179–195.
[20] V. E. Tarasov, G.M. Zaslavsky, *Physica A* **368**(2006) 399.
[21] A. Tofighi, H .Nasrolahpour, *Physica A* **374**(2007) 41.
[22] H.-P. Breuer and F. Petruccione, The Theory of Open Quantum Systems (*Oxford Univ. Press, Oxford*, 2007).
   C.W. Lai, P. Maletinsky, A. Badolato, and A. Imamoglu, *Phys. Rev. Lett.* **96**(2006)167403
   D. Aharonov, A. Kitaev, and J. Preskill, *Phys. Rev. Lett.* **96** (2006)050504.
   Á. Rivas, S. F. Huelga, M. B. Plenio, *Phys. Rev. Lett.* **105**(2010) 050403.
   J. Shao , J. Chem. Phys. 120(2004) 5053.
[23] M. A. Abramowitz , I. A. Stegun, (Eds.), Handbook of Mathematical Functions with Formulas, Graphs, and Mathematical Tables (*Dover*, New York, 1972).
[24] P. Maass, F. Scheffler, *Physica A* **314** (2002) 200 – 207.
[25] P. Zavada, *J.Appl.Math.* **2** (2002) 163-197.
[26] A. Raspini, *Phys. Scr.* **64** (2001)20.
[27] E. Goldfain, *Chaos, Solitons & Fractals* **28** (2006) 913–922.
[28] E. Goldfain, *Comm. Non. Sci. Num. Siml.* , **13**(2008) 1397-1404.

[29] E. Goldfain, *Comm. in Nonlin. Dynamics and Numer. Simulation*, **14** (2009) 1431-1438.
[30] R. Herrmann, *Phys. Lett. A* **372**(2008)5515-5522.
[31] S. I. Muslih, Om P. Agrawal, D. Baleanu, *J. Phys. A: Math. Theor.* **43** (2010) 055203.
[32] R. A. El-Nabulsi , *Chaos, Solitons & Fractals* **42** (2009) 2614–2622.
   R. A. El-Nabulsi, *Chaos, Solitons & Fractals* **41** (2009) 2262–2270.
   R. A. El-Nabulsi, *Int. J. Mod. Geom. Meth. Mod. Phys.* **5**(2008)863.

Article

Plane Wave Solutions of Weakened Field Equations in a Plane Symmetric Space-time-II

Sanjay R. Bhoyar[*] & A. G. Deshmukh[&]

[&]Ex. Head, Department of Mathematics, Govt. Vidarbha Institute of Science and Humanities, Amravati, (India), Joint Director, Higher Education, Nagpur Division, Nagpur-440022 (India).
[*]Dept. of Math., Gopikabai Sitaram Gawande College, Umerkhed- 445206, India

Abstract

In this paper, we propose to obtain the $Z = (t/z)$ – type plane gravitational waves in a set of five vacuum weakened equations in the space-time introduced by us having plane symmetry in the sense of Taub [**Ann.**Math.**53**,472 (1951)]. These vacuum field equations has been suggested as alternatives to the Einstein vacuum field equations of general relativity. Furthermore the physical significance of modified gravitational waves for the space-time is obtained.

Keywords: plane symmetry, plane gravitational waves, curvature tensor, Ricci tensor, weakened field equations.

1. Introduction

The theory of plane gravitational waves in general relativity has been introduced by many investigators like Einstein and Rosen (1937); Bondi, Pirani and Robinson (1959); Takeno (1961). Takeno (1961) has discussed the mathematical theory of plane gravitational waves and classified them into two categories, namely, $(z-t)$ and $Z = (t/z)$–type wave according as the phase function $Z = (z-t)$ and $Z = (t/z)$–type wave respectively. According to him, a plane wave g_{ij} is a non-flat solution of Ricci tensor $R_{ij} = 0$ in general relativity and in some suitable coordinate system; all the components of the metric tensor are functions of a single variable $Z = Z(z,t)$ (i.e. a phase function). Takeno (1957) deduce the space-time having plane symmetry characterized by Taub (1951) for $Z = (z-t)$-type plane gravitational waves as

$$ds^2 = -A(dx^2 + dy^2) - C(dz^2 - dt^2) \qquad (1.1)$$

---

[*] Correspondence Author: Prof. Sanjay R. Bhoyar, Department of Mathematics, College of Agriculture, Darwha -445 202, India. E-mail: sbhoyar68@yahoo.com.

where A and C are arbitrary functions of Z, $Z = z - t$.

Recently Bhoyar et al. (2011) transform the metric (1.1) to (1.2) using suitable transformations for $Z = (t/z)$ – type plane gravitational waves, which take the form

$$ds^2 = -A(dx^2 + dy^2) - Z^2 C dz^2 + B dt^2 \qquad (1.2)$$

Where A and B are functions of Z, $Z \equiv (t/z)$.

Lovelock (1967a, b) has considered a set of five weakened field equations (WFE) in vacuum, namely,

$$R^h_{ijk;h} = 0, \qquad (1.3)$$

$$(-g)^{\frac{1}{4}}\left[g^{ih}R_{kj;ih} - g^{ih}R_{ij;kh} + \frac{1}{6}R_{;kj} - \frac{1}{6}g_{jk}g^{ih}R_{;ih} - R^{ih}C_{jhik} + \frac{R}{6}g^{ih}C_{jhik}\right] = 0, \qquad (1.4)$$

$$(-g)^{\frac{1}{2}}\left[g^{hj}g^{ki}(2R_{jlim}R^{ml} + g^{ml}R_{ij;lm} - R_{;ij}) - \frac{1}{2}g^{hk}(R^l_m R^m_l - g^{lm}R_{;lm})\right] = 0, \qquad (1.5)$$

$$(-g)^{\frac{1}{2}}\left[(g^{hk}g^{rs} - \frac{1}{2}g^{hr}g^{ks} - \frac{1}{2}g^{hs}g^{kr})R_{;rs} + R(R^{kh} - \frac{1}{4}g^{kh}R)\right] = 0, \qquad (1.6)$$

and $\qquad R^{ij}_{;k} = 0, \qquad (1.7)$

where a semicolon (;) followed by an index denotes covariant differentiation and C_{jhik} is the Weyl curvature tensor defined by

$$C_{jhik} = R_{jhik} - \frac{1}{2}(R_{ji}g_{hk} - R_{hi}g_{jk} - R_{jk}g_{hi} + R_{hk}g_{ij}) + \frac{R}{6}(g_{ij}g_{hk} - g_{hi}g_{jk}). \qquad (1.8)$$

Kilmister and Newman (1961) have originally proposed the vacuum weakened field equations (1.2)-(1.6). These field equations are suggested as various alternatives to the Einstein field equation of general relativity in vacuum. The Einstein vacuum field equation of general relativity is given by

$$R_{ij} = 0. \qquad (1.9)$$

The solution of (1.3)) together with the trajectories of test particles (geodesics hypothesis) give agreement with experiment. Lovelock (1967a, b) obtained the solutions of WFE field equations (1.2)-(1.6) in a spherically symmetric space-time and he proved to be gravitationally unphysical metric by geodesics hypothesis in the sense that these solutions correspond to the static situation of an isolated mass at origin which repels the test particles.

Consequently the physical aspects of weakened field equations are not well established through many researchers (for examples: Thompson 1963; Kilmister 1966; Rund 1967; Lovelock 1967a, b; Swami 1970; Lal and Singh 1973; Lal and Pandey 1975; Pandey 1975) but have tried to investigate the solutions to interpret the useful results. Thompson (1963) made detailed study of these field equations and concluded that they are too weak.

The various alternative vacuum weakened field equations are weaker than the Einstein vacuum field equations in the sense that they each admit (1.4) as a solutions and hence they have been called WFE. Swami (1970) has solved three solutions of the weakened field equations $R_{ij;k} - R_{ik;j} = 0$ with $R_{ij} \neq 0$, $R_{ij} \neq \lambda g_{ij}$ and has discussed the geometrical and dynamical properties of these solutions.

Furthermore R^h_{ijk} satisfies the Bianchi identities

$$R^h_{ijk;m} + R^h_{ikm;j} + R^h_{imj;k} = 0 . \qquad (1.10)$$

From which

$$R^i_{j;i} = \frac{1}{2} R_{;j} , \qquad (1.11)$$

where $R^i_j = g^{ih} R_{hj}$.

In this paper, we study the $Z = (t/z)$-type plane gravitational waves of vacuum weakened field equations (1.3)-(1.7) in the metric (1.2). In addition to this, we assume the metric (1.2) has non-conformally flat. Physical significance of modified gravitational waves for metric (1.2) is obtained. We have obtained some useful results in the form of theorems under curvature properties with conclusions.

2. Plane Symmetric metric and curvature properties

The components of contravariant tensor g^{ij} from the metric (1.2) are

$$g^{11} = g^{22} = -\frac{1}{A}, \quad g^{33} = -\frac{1}{Z^2 C}, \text{ and } \quad g^{44} = \frac{1}{C} \qquad (2.1)$$

The non vanishing components of Christoffel symbols are

$$Z\begin{Bmatrix}3\\11\end{Bmatrix} = Z\begin{Bmatrix}3\\22\end{Bmatrix} = \begin{Bmatrix}4\\11\end{Bmatrix} = \begin{Bmatrix}4\\22\end{Bmatrix} = \frac{\overline{A}}{2Cz},$$

$$\frac{1}{Z}\begin{Bmatrix}1\\13\end{Bmatrix} = \frac{1}{Z}\begin{Bmatrix}2\\23\end{Bmatrix} = -\begin{Bmatrix}1\\14\end{Bmatrix} = -\begin{Bmatrix}2\\24\end{Bmatrix} = -\frac{\overline{A}}{2Az},$$

$$\begin{Bmatrix}3\\33\end{Bmatrix} = -Z\begin{Bmatrix}3\\34\end{Bmatrix} = -\frac{1}{Z}\begin{Bmatrix}4\\33\end{Bmatrix} = -\frac{(2C + \overline{C}Z)}{2Cz},$$

$$\frac{1}{Z}\begin{Bmatrix}4\\34\end{Bmatrix} = -\begin{Bmatrix}4\\44\end{Bmatrix} = Z\begin{Bmatrix}3\\44\end{Bmatrix} = -\frac{\overline{C}}{2Cz} \qquad (2.2)$$

Using (1.2), (2.1) and (2.2), the components of curvature tensor R_{ijkl} are as follows:

$$\frac{R_{1313}}{Z^2} = \frac{R_{2323}}{Z^2} = -\frac{R_{2324}}{Z} = -\frac{R_{1314}}{Z} = R_{1414} = R_{2424} = \psi, \qquad (2.3)$$

The components of covariant and contra-variant Ricci tensor, from (2.1) and (2.2) are as follows:

$$\frac{R_{33}}{Z^2} = -\frac{R_{34}}{Z} = R_{44} = \frac{\psi}{z^2}, \text{ (Say)}, \qquad (2.4a)$$

$$Z^2 R^{33} = ZR^{34} = R^{44} = \frac{\psi}{C^2 z^2} \text{ and all other } R_{ij} = R^{ij} = 0, \qquad (2.4b)$$

where

$$\psi = \psi(Z) = \frac{\overline{\overline{A}}}{A} - \frac{\overline{A}^2}{2A^2} - \frac{\overline{A}\overline{C}}{AC}. \qquad (2.5)$$

From (2.2), (2.4b), we obtain

$$\frac{1}{Z^4} R_{33;33} = -\frac{1}{Z^3} R_{33;34} = \frac{1}{Z^2} R_{33;44} = -\frac{1}{Z^3} R_{34;33} = \frac{1}{Z^2} R_{34;34} = -\frac{1}{Z} R_{34;44}$$

$$= \frac{1}{Z^2} R_{44;33} = -\frac{1}{Z} R_{44;34} = R_{44;44} = K \quad (2.6)$$

where $\quad K = \dfrac{1}{Z^4}\left(\overline{\overline{\psi}} - \dfrac{2\overline{C}\overline{\psi}}{C} - \dfrac{5\overline{C}\overline{\psi}}{C} + \dfrac{8\overline{C}^2 \psi}{C^2}\right).$ (2.7)

Here a bar (–) overhead letter denotes the differentiation with respect to Z (i.e. $\overline{\psi} = \dfrac{\partial \psi}{\partial Z}$ and $\overline{\overline{\psi}} = \dfrac{\partial^2 \psi}{\partial Z^2}$).

Also for metric (1.2), we deduce, from (2.4 a,b), that

a) The scalar curvature R defined by $R = g^{ij} R_{ij}$ is zero i.e. $R = 0$,

b) $g = \det(g_{ij}) = -Z^2 A^2 C^2$,

c) $R_m^l R_l^m = R_{im} R^{im} = 0$

and d) $R_{jlim} R^{ml} = 0$. (2.8)

3. Solutions of weakened field equations (1.3)-(1.7)

By its very structure the metric (1.1) is non-conformally flat which implies that Weyl curvature tensor (1.2) in view of (2.8a) (i.e. $R = 0$) reduces to

$$C_{jhik} = R_{jhik} - \frac{1}{2}(R_{ji} g_{hk} - R_{hi} g_{jk} - R_{jk} g_{hi} + R_{hk} g_{ij}). \quad (3.1)$$

From Bianchi identities (1.10), we find

$$R^h_{ijk;h} = R_{ij;k} - R_{ik;j} \quad (3.2)$$

where $\quad R_{ij;k} = R_{ij,k} - \Gamma^m_{jk} R_{im} - \Gamma^m_{ik} R_{mj}.$

On substituting the components of $\begin{Bmatrix} i \\ jk \end{Bmatrix}$ and R_{ij} from (2.2) and (2.4) respectively in R.H.S. of equation (3.2), it is seen that

$$R_{ij;k} - R_{ik;j} = 0, \tag{3.3}$$

or equivalently, from (3.2),

$$R^h_{ijk;h} = 0, \tag{3.4}$$

which is weakened field equation (1.3).

Also it follows from (3.3) that

$$R_{ij;kh} = R_{ik;jh}. \tag{3.5}$$

In view of equations (2.8a) and (3.5), the weakened field equation (1.4) reduces to

$$(-g)^{\frac{1}{4}} R^{ih} C_{jhik} = 0. \tag{3.6}$$

Using (3.1) in (3.6), we obtain

$$(-g)^{\frac{1}{4}} \left[R^{ih} R_{jhik} - \frac{1}{2} (R_{ji} R^i_k - R^{ih} R_{hi} g_{jk} - R^i_i R_{jk} + R^h_j R_{hk}) \right] = 0, \tag{3.7a}$$

which reduces to

$$(-g)^{\frac{1}{4}} \left[R^{ih} R_{jhik} - \frac{1}{2} (g_{jk} R^k_i R^i_k - g_{jk} R^{ih} R_{hi} - R R_{jk} + g_{jk} R^j_h R^h_j) \right] = 0. \tag{3.7b}$$

Using (2.8a, c, d) in (3.7b), it is seen that no term remains in LHS and hence (3.7b) i.e. (3.6) is identically satisfied.

It is observed that the weakened field equation (1.5) is satisfied by $R = 0$ (i.e. 2.8a) alone. Hence the following theorem:

Theorem 1: *The g_{ij} given by (1.2) is a solution of weakened field equation (1.3), (1.4) and (1.6).*

Theorem 2: *A necessary and sufficient condition that g_{ij} given by (1.2) be a solutions of WFE (1.5) are i) $\psi = 0$, ii) $K = 0$, where*

$$\psi = \frac{z^2 R_{33}}{Z^2} \quad \text{and} \quad K = \frac{1}{z^4}\left(\overline{\overline{\psi}} - \frac{2\overline{\overline{C}}\psi}{C} - \frac{5\overline{C}\overline{\psi}}{C} + \frac{8\overline{C}^2\psi}{C^2}\right).$$

Proof:

First, let g_{ij} given by (1.2) be the solution of WFE (1.5).

By using equation (2.8), the WFE (1.5) reduces to

$$(-g)^{1/2} g^{hj} g^{ki} g^{ml} R_{ij;lm} = 0. \tag{3.8}$$

Case i) $\psi = 0$, the result is obvious.

Case ii) The above equation (3.8) is identically satisfied for all the values of h, k except for $h, k = 3, 4$. On simplification, for $h, k = 3, 4$, by the virtue of (2.4) and (2.6), equation (3.8) gives,

$$K = \frac{1}{z^4}\left(\overline{\overline{\psi}} - \frac{2\overline{\overline{C}}\psi}{C} - \frac{5\overline{C}\overline{\psi}}{C} + \frac{8\overline{C}^2\psi}{C^2}\right).$$

Conversely, if

$$K = \frac{1}{z^4}\left(\overline{\overline{\psi}} - \frac{2\overline{\overline{C}}\psi}{C} - \frac{5\overline{C}\overline{\psi}}{C} + \frac{8\overline{C}^2\psi}{C^2}\right) = 0.$$

By (2.4) and (2.6), we have

$$(-g)^{1/2} g^{hj} g^{ki} g^{ml} R_{ij;lm} = 0.$$

Introducing the result (2.8) and above equation we get,

L.H.S. of (1.5) = 0.

So WFE (1.5) is identically satisfied.

Theorem 3: *A necessary and sufficient condition that g_{ij} given by (1.2) is the solution of WFE (1.7) is* $\overline{\psi} = \frac{2\psi\overline{C}}{C}$.

Proof:

Let g_{ij} given by (1.2) be a solution of WFE (1.7).

By the definition of covariant derivative, equation (1.7)

$$\frac{\partial R^{ij}}{\partial x^k} + \Gamma^i_{pk} R^{pj} + \Gamma^j_{pk} R^{ip} = 0. \qquad (3.9)$$

By using the components of Ricci tensors and Christoffel's symbols, the equation (3.9) is identically satisfied for all values of i, j, k except when $i, j, k = 3, 4$,

i.e. $\quad \overline{\psi} = \dfrac{2\psi \overline{C}}{C}.$

Conversely

If $\overline{\psi} = \dfrac{2\psi \overline{C}}{C}$, then it is seen that the WFE (1.7) is identically satisfied,

i.e. $\quad R^{ij}_{;k} = 0.$

This implies g_{ij} be the solution of WFE (1.7). Hence the theorem.

4. Physical significance of modified gravitational waves for the space-time metric

To attempt the physical significance of modified gravitational waves for the space-time metric (5.1.1), it is useful to consider the effect of motion of test particles introduced into the system. We assume the motion of test particles in a curved geometry which describe the geodesics. The equations of motion of such particles are given by

$$\frac{d^2 x^k}{ds^2} + \Gamma^k_{ij} \frac{dx^i}{ds} \frac{dx^j}{ds} = 0$$

or equivalently

$$\frac{d}{ds}\left(g_{kl} \frac{dx^l}{ds}\right) - \frac{1}{2} g_{ij,k} \frac{dx^i}{ds} \frac{dx^j}{ds} = 0. \qquad (4.1)$$

The metric (1.2) represents a modified gravitational waves i.e. generalized plane gravitational waves. For the metric (1.2), first two integrals of motion from (4.1) gives

$$A\frac{dx}{ds} = \lambda \qquad (4.2)$$

and $\quad A\dfrac{dy}{ds} = \mu$, $\qquad (4.3)$

where λ, μ are constants. For $k = 3, 4$ the equation (4.1) yields

$$\left(\frac{1}{Z}\right)\frac{d}{ds}\left[\frac{dz}{ds}(Z^2 C)\right] = \frac{d}{ds}\left[\frac{dt}{ds}(C)\right]. \qquad (4.4)$$

Noting that $Z = t/z$, we take

$$\frac{dZ}{ds} = \gamma = \text{constant on geodesics.}$$

From equations (4.2) and (4.3), we obtain

$$\int \frac{A}{m} dZ = \frac{(\gamma x - k_1)}{\lambda} = \frac{(\gamma y - k_2)}{\mu} \qquad (4.5)$$

where k_1 and k_2 are constants of integration and $m = A^2$.

Equation (1.1) can be put in another form as

$$\gamma^2 z \left[Cz + 2CZ\frac{dz}{dZ}\right] = 1 + \frac{A}{m}(\lambda^2 + \mu^2) \qquad (4.6)$$

Which on integrating and using equations (4.2) and (4.3) reduces to

$$\gamma^2 t \int C dz = Z + \gamma(\lambda x + \mu y) - (\lambda k_1 + \mu k_2) \qquad (4.7)$$

To ensure the plane character of waves we may consider that Z is a function of x^i (Takeno 1961). Consequently C may be assumed as the linear function of Z. Integrating (4.7) and noting $Z = (t/z)$ we get,

$$\gamma^2 (\tau t z + \xi z^2) + \gamma(\lambda x + \mu y) + Z = \delta, \qquad (4.8)$$

where τ, ξ and δ are constants. The right hand side constant can be made zero by a suitable choice of the origin without loss of generality.

Discussions

1] If the test particle is at rest (i.e. at rest means the coordinates of the particle do not change) at the origin i.e. (x, y, z) = (0, 0, 0) $\Rightarrow \delta = 0$ at $t = 0$, then equation reflects that either $t = 0$ or γ is real only in the negative direction of z, hence no other particle with real γ passing through origin will return to the resting particle. Further in case of moving particle no return can take place.

2] For the space-time metric (1.2), the coordinates along the world line of test particles are geodesic. If the curvature of trajectories of test particles is everywhere zero, then the geodesics are straight line. If the particle released from rest in z, t plane, then, for (1.2), world line of such resting particle bring into a world line of free particle moving in the plane. Hence a particle starting from rest in z, t plane must necessarily move along the geodesic and the solutions of geodesic equations are unique

3] If two particle perform the general motion then we have to consider two equations of type (4.7) with constants γ and γ' and with $\delta = \delta' = 0$. This leads to $\gamma = \gamma'$. Taking $\tau', \xi', \lambda', \mu'$ as corresponding constants for second particle, we find

$$\gamma = \frac{(\lambda' - \lambda)x + (\mu' - \mu)y}{(\tau - \tau')tz + (\xi - \xi')z^2},$$

as the condition for consecutive meeting of particle.

Conclusions

In this paper:

1] The plane wave g_{ij} given by metric (1.2) representing a non-conformal flat space-time with the scalar curvature zero ($R = 0$) is a solution of the weakened field equation (1.6). It is also a solution of WFE (1.3), (1.4) and (1.5) under the curvature properties (2.8). In non-conformal flat space-time (1.2), the solution of (1.6) follows from Bianchi identity gives (1.10) and (3.2). When $R = 0$, then $R_{;j} = 0$ and (1.6) implies that $R^i_{j;i} = 0$.

2] The physical significance of modified gravitational waves for the space-time is obtained on the basis of geodesics hypothesis. It is observed that the effects of gravitational field can be changed by purely gravitational effects associated with motion of free particle in a curved space-time.

4] It is observed that the metric (1.2) is a non-flat and it is an exact solution of $R_{ij}=0$ if and only if $K=0$. In such case, plane wave metric satisfied the curvature properties given in (2.8). Hence we have the following results:

a) The g_{ij} given by above Takeno's plane wave metric is a solution of weakened field equations (1.3), (1.4) and (1.6).

b) The g_{ij} given by above Takeno's plane wave metric is a solution of weakened field equation (1.5) if and only if $\overline{\overline{\psi}}=0$, where $\psi = \dfrac{z^2 R_{33}}{Z^2}$.

c) The g_{ij} given by above Takeno's plane wave metric is a solution of weakened field equation (1.6) if and only if $\overline{\psi} = \dfrac{2\psi \overline{C}}{C}$.

References

[1] H.Takeno (1961): The mathematical theory of plane gravitational waves in general relativity, *Sci.Rep.Inst.Theo.Phy.Hiroshima University, Japan.*

[2] D.Lovelock (1967): Weakened Field Equations in general relativity admitting an "unphysical" metric, *Commun. Math. Phys., 5, 264-214.*

[3] S.N.Pandey (1975): Plane wave solutions of Weakened Field Equations in Peres space-time, *Tensor N.S., 29, 264.*

[4] Lal K.B and Pandey S.N (1975): On Plane wave-like solutions of Weakened Field Equations in general relativity, *Tensor N.S., 29, 297-298.*

[5] Chirde.V, Metkar .A and Deshmukh A. (2006): $(z-t)$-Type Plane wave solutions of Weakened Field Equations, *Tensor N.S., 29, 264.*

[6] Rane.R.S and Katore S. D (2009): Generalized Plane gravitational waves of Weakened Field Equations general relativity, *J.Math.Phys.50.053504 (2009).*

[7] Bhoyar S.R and Deshmukh (2011): The Z=Z (t/z)-type Plane wave solutions of the field equations of general relativity in a Plane symmetric space-time. *Int. Jour .of Theo. Phy. DOI 10.1007/s10773-011-0853-4.*

Plane Wave Solutions of Field Equations of Israel and Trollope's Unified Field Theory in V_5

Gowardhan P. Urkude[&], J. K. Jumale[*] & K. D. Thengane[@]

[&]Department of Physics, Govt. Resi. Women's Polytechnic, Yavatmal, (M. S.), India
[*]Department of Physics, R. S. Bidkar College, Hinganghat, Wardha, India
[@]Principal, N.S. Science & Arts College, Bhadravati, Dist. Chandrapur, India

Abstract

Israel and Trollope (1961) have developed field equations in the form of two sets in their non-symmetric unified field theory. Shrivastava (1974) and Pradhan (1977) have obtained wave solutions of the field equations of Israel and Trollope in four-dimension using generalized Takeno's space-time and generalized Peres space-time respectively. In the paper [3], recently we have investigated $(z-t)$-type plane wave solutions of Israel and Trollope's field equations in five-dimensional space-time. In the present paper, we have obtained the five dimensional plane wave solutions of field equations of Israel and Trollope's unified field theory in the space-time of Ambatkar (2002) for the function $Z = t/z$. It is found that, the second set of the field equations of this unified field theory does not have any plane wave solution.

Keywords: field equation, plane wave solution, unified field theory, Israel, Trollope.

1. Introduction

Takeno (1961) has solved field equations of Einstein's general relativity as well as non-symmetric unified field theory and obtained plane wave solutions in four-dimensional space-time which lead to $(z-t)$-type and (t/z)-type plane waves. A new set of field equations, alternative to Einstein's unified field equations, as proposed by Israel and Trollope (1961) has been considered by Srivastava (1974) in the two types of generalized Takneo's space-times one proposed by Srivastava (1974) himself and other suggested by the Lal and Ali (1970) and found wave solutions in four dimensional space-time in the sense that plane wave solutions in the metric of Takeno (1961) are obtainable from the various solutions obtained by Srivastava (1974) in the two generalized metrics. The plane wave-like solutions in Peres space-time of the field equations of Israel and Trollope (1961) have also been derived by Srivastava (1974). Furthermore, plane wave like solutions of field equations of Israel and Trollope (1961) in the generalized Peres space-time have been investigated by Pradhan (1977).

[*] Correspondence Author: J. K. Jumale, Department of Physics, R. S. Bidkar College, Hinganghat - 442 301 Wardha, India. E-mail: jyotsnajumale@yahoo.com

In a past few years there have been many attempts to construct a unified field theory based on the idea of multidimensional space-time. Most recent efforts have been diverted at studying theories in which the dimensions of the space-time are greater than (3+1) of the order which we observe. The idea that space-time should be extended from four to higher five dimension was introduced by Kaluza and Klein (1921,1926) to unify gravity and electromagnetism. Wesson (1983,1984) and Reddy D.R.K. (1999) have studied several aspects of five dimensional space-time in variable mass theory and bi-metric theory of relativity respectively. Inspired by work in string theory and other field theories, there has been a considerable interest in recent times to find solutions of the Einstein's field equations in dimensions greater than four. In the present work we would like to find plane wave solutions in the frameworks of five-dimensional space-time. Many authors have extended Takeno's work in general as well as non-symmetric unified field theory proposed by Einstein to higher dimensions using the different space-times [Thengane (2000), Ladke(2004), Jumale (2006)]. With this motivation the study regarding $(z-t)$-type plane wave solutions of the field equations of Israel and Trollope has been carried out to higher five dimensional space time in our earlier paper [3]. Recently Warade (2006) has extended the work regarding (t/z)-type plane wave solutions of the field equations of Einstein's non-symmetric unified field theory to higher five dimensional space-time on the lines of Takeno (1961). We observed that (t/z)-type plane wave solutions obtained by Warade (2006) in ENSUFT can further be studied to the case of Israel and Trollope's unified field theory and therefore, an attempt has been made in the present paper. Therefore, in this paper, we have solved the field equations of Israel and Trollope's unified field theory and investigated plane wave solutions in five dimensional space-time of Ambatkar (2002) for the function $Z = t/z$. To find out plane wave solutions in V_5, we consider the first set [I] of five dimensional field equations of Israel and Trollope's unified field theory as under

$$\theta^{ij}{}_{;\mu}{}^* = 0, \qquad (i, j = 1,2,3,4,5) \tag{1.1}$$

$$\Gamma_i^* = \Gamma_{[ij]}^{*j} = 0, \tag{1.2}$$

$$\theta^{[ij]}{}_{,j} = 0, \tag{1.3}$$

$$R_{(ij)} = \alpha M_{ij}(R_{[\,]}), \tag{1.4}$$

$$R = 0 \tag{1.5}$$

while the second set [II] of five dimensional field equations consists of

$$\theta^{ij}{}_{;\mu}{}^* = 0, \tag{1.6}$$

$$\Gamma_i^* = \Gamma_{[ij]}^{*j} = 0, \qquad (1.7)$$

$$\theta^{[ij]}{}_{,j} = 0, \qquad (1.8)$$

$$bR(R_{(ij)} - \frac{1}{4} R g_{ij}) = \alpha M_{ij}(R_{[\,]}), \qquad (1.9)$$

where R be a non-vanishing constant.

Israel and Trollope have assumed $bR = 1$ therefore, the field equation (1.9) replaced by

$$R_{(ij)} - \frac{1}{4b} g_{ij} = \alpha M_{ij}(R_{[\,]}), \qquad (1.10)$$

where α and b are constants and θ^{ij} is a contra-variant tensor density defined by

$$\theta^{ij} = \sqrt{g}(g^{ij} + \alpha R^{[ij]}) \qquad (1.11)$$

and $M_{ij}(R_{[\,]})$, is known as the Maxwellian of the tensor $R_{[ij]}$ which is defined by

$$M_{ij}(R_{[\,]}) = (1/4) g_{ij}(R_{[lm]} R^{[lm]}) - R_{[il]} R_{[km]} g^{lm}. \qquad (1.12)$$

In both the two sets of field equations a semicolon (;) followed by asterisk $*$ denotes covariant differentiation with respect to connections Γ_{jk}^{*i}. The Ricci tensor R_{ij} is defined by

$$R_{ij} \equiv \Gamma_{il,j}^{l} - \Gamma_{ij,l}^{l} + \Gamma_{tj}^{l}\Gamma_{il}^{t} - \Gamma_{tl}^{l}\Gamma_{ij}^{t}. \qquad (1.13)$$

In this paper, we propose to obtain the (t/z)-type plane wave solutions of field equations [I] and (II) of Israel and Trollope in five dimensional space-time of Ambatkar (2002). The paper is organized as under : section 2 contains components of contra-variant tensor density. In the section 3, we have explained components of the tensors $g^{ij}, R^{[ij]}, g_{ij}, R_{[ij]}$ and $M_{ij}(R_{[\,]})$. Section 4 is dealt with solutions of the field equations (1.1), (1.2) and (1.3). Solutions of the field equations (1.4) and (1.5) are obtained in the fifth section. Section 6 is devoted to the study of plane wave solutions of second set of field equations of Israel and Trollope's unified field theory and in the last section of the present paper, we summarize and conclude the results.

2. Components of contra-variant tensor density θ^{ij} in V_5

Recently, Warade (2006) has extended the work of Takeno (1961) regarding plane wave solutions to higher five dimensional space-time in Einstein's non-symmetric unified field theory with the space-time of Ambatkar (2002). The solutions are obtained by Warade (2006) in five dimension are in the format of Takeno (1961). Thus in [2], Warade (2006) has obtained the non-symmetric tensor g_{ij} as a plane wave solutions of field equations of ENSUFT in the five dimensional space-time of Ambatkar (2002) :

$$ds^2 = -Adx^2 - 2Ddxdy - Bdy^2 - 2Fdxdu - 2Gdydu - Hdu^2 - Z^2(C-E)dz^2 - 2ZEdzdt$$

$$+ (C+E)dt^2 \quad (2.1)$$

such that $g_{ij} = \begin{bmatrix} -A & -D & -F & -Z\sigma_1/z & Z\sigma_1/t \\ -D & -B & -G & -Z\rho_1/z & Z\rho_1/t \\ -F & -G & -H & -Z\eta_1/z & Z\eta_1/t \\ Z\sigma_1/z & Z\rho_1/z & Z\eta_1/z & -Z^2(C-E) & -ZE \\ -Z\sigma_1/t & -Z\rho_1/t & -Z\eta_1/t & -ZE & C+E \end{bmatrix}$ (2.2)

where $A, B, D, F, G, H, \sigma_1, \rho_1$ and η_1 are functions of $Z \equiv Z(t/z)$.

To solve the five dimensional field equations of Israel and Trollope's unified field theory, we have in need to find components of contra-variant tensor density θ^{ij} in V_5 and therefore, we consider the non-symmetric covariant tensor S_{ij} in the same form of g_{ij} as given in (2.2).

$$S_{ij} = \begin{bmatrix} -A & -D & -F & -Z\sigma_1/z & Z\sigma_1/t \\ -D & -B & -G & -Z\rho_1/z & Z\rho_1/t \\ -F & -G & -H & -Z\eta_1/z & Z\eta_1/t \\ Z\sigma_1/z & Z\rho_1/z & Z\eta_1/z & -Z^2(C-E) & -ZE \\ -Z\sigma_1/t & -Z\rho_1/t & -Z\eta_1/t & -ZE & C+E \end{bmatrix} \quad (2.3)$$

which implies that

$$\det.(S_{ij}) = g = mn > 0, \quad m = -ABH + AG^2 + HD^2 + BF^2 - 2DFG \text{ and } n = -Z^2C^2. \quad (2.4)$$

It is to be noted that the symmetric part of S_{ij} corresponds to the metric tensor of the space time (2.1).

According to Israel and Trollope (1961), we have a relation between covariant tensor S_{ij} and contra-variant tensor density θ^{ij} such that

$$S_{i\alpha}\theta^{j\alpha} = \delta_i^j (\det \theta^{\nu\mu})^{1/2} = \delta_i^j (\det S_{ij})^{1/2} = \delta_i^j \sqrt{g} \tag{2.5}$$

where $\det(\theta^{\nu\mu}) = \det(S_{ij})$. $\hspace{2cm}$ (2.6)

Putting the values of S_{ij} from (2.3) in the equation (2.5), we get twenty five equations and after solving them we have the component of θ^{ij} as follows:

$$\theta^{ij} = \begin{bmatrix} \dfrac{(BH-G^2)\sqrt{g}}{m} & \dfrac{(FG-DH)\sqrt{g}}{m} & \dfrac{(DG--BF)\sqrt{g}}{m} & -\dfrac{T}{zZ} & -\dfrac{T}{z} \\ \dfrac{(FG-DH)\sqrt{g}}{m} & \dfrac{(AH-F^2)\sqrt{g}}{m} & \dfrac{(DF-AG)\sqrt{g}}{m} & -\dfrac{U}{zZ} & -\dfrac{U}{z} \\ \dfrac{(DG--BF)\sqrt{g}}{m} & \dfrac{(DF-AG)\sqrt{g}}{m} & \dfrac{(AB-D^2)\sqrt{g}}{m} & -\dfrac{V}{zZ} & -\dfrac{V}{z} \\ \dfrac{T}{zZ} & \dfrac{U}{zZ} & \dfrac{V}{zZ} & \dfrac{1}{Z^2}(\omega-\dfrac{\sqrt{g}}{C}) & \dfrac{\omega}{Z} \\ \dfrac{T}{z} & \dfrac{U}{z} & \dfrac{V}{z} & \dfrac{\omega}{Z} & (\omega+\dfrac{\sqrt{g}}{C}) \end{bmatrix} \tag{2.7}$$

where $T = \dfrac{-\sqrt{g}}{mC}[\sigma_1(BH-G^2) + \rho_1(FG-DH) + \eta_1(DG-BF)]$,

$$U = \dfrac{-\sqrt{g}}{mC}[\sigma_1(FG-DH) + \rho_1(AH-F^2) + \eta_1(DF-AG)],$$

$$V = \dfrac{-\sqrt{g}}{mC}[\sigma_1(DG-BF) + \rho_1(DF-AG) + \eta_1(AB-D^2)],$$

$$\omega = \dfrac{1}{C}[\dfrac{1}{z^2}(\sigma_1 T + \rho_1 U + \eta_1 V) - \dfrac{E\sqrt{g}}{C}]$$

$$= -\dfrac{\sqrt{g}}{z^2 mC^2}\{[\sigma_1^2(BH-G^2) + \rho_1^2(AH-F^2) + \eta_1^2(AB-D^2)] + 2[\sigma_1\rho_1(FG-DH)$$

$$+ \sigma_1\eta_1(DG-BF) + \rho_1\eta_1(DF-AG)]\} - \dfrac{E\sqrt{g}}{C^2}. \tag{2.8}$$

3. Components of the tensors $g^{ij}, R^{[ij]}, g_{ij}, R_{[ij]}$ and $M_{ij}(R_{[\ ]})$ in V_5

In this section we find the components of tensors $g^{ij}, R^{[ij]}, g_{ij}, R_{[ij]}$ and $M_{ij}(R_{[\ ]})$. The components of g^{ij} and non-vanishing components of $R^{[ij]}$ can be obtained by substituting the values of tensor density θ^{ij} into the equation (1.11) as under

$$g^{ij} = \begin{bmatrix} \frac{BH-G^2}{m} & \frac{FG-DH}{m} & \frac{DG-BF}{m} & 0 & 0 \\ \frac{FG-DH}{m} & \frac{AH-F^2}{m} & \frac{DF-AG}{m} & 0 & 0 \\ \frac{DG-BF}{m} & \frac{DF-AG}{m} & \frac{AB-D^2}{m} & 0 & 0 \\ 0 & 0 & 0 & \frac{1}{Z^2}(\omega'-\frac{1}{C}) & \frac{\omega'}{Z} \\ 0 & 0 & 0 & \frac{\omega'}{Z} & (\omega'+\frac{1}{C}) \end{bmatrix} \quad (3.1)$$

where $\omega' = \omega/\sqrt{g} = \frac{1}{c\sqrt{g}}[t^2(\sigma_1 T + \rho_1 U + \eta_1 V) - \frac{E\sqrt{g}}{C}]$

$$= -\frac{1}{z^2 mC^2}\{[\sigma_1^2(BH-G^2) + \rho_1^2(AH-F^2) + \eta_1^2(AB-D^2)] + 2[\sigma_1\rho_1(FG-DH)$$

$$+ \sigma_1\eta_1(DG-BF) + \rho_1\eta_1(DF-AG)]\} - (E\sqrt{g}/C^2). \quad (3.2)$$

and $\quad ZR^{[14]} = -ZR^{[41]} = R^{[15]} = -R^{[51]} = -T/z\alpha\sqrt{g}$,

$\quad ZR^{[24]} = -ZR^{[42]} = R^{[25]} = -R^{[52]} = -U/z\alpha\sqrt{g}$,

$\quad ZR^{[34]} = -ZR^{[43]} = R^{[25]} = -R^{[52]} = -V/z\alpha\sqrt{g}$. $\quad (3.3)$

From (3.1) the components of g_{ij} are

$$g_{ij} = \begin{bmatrix} -A & -D & -F & 0 & 0 \\ -D & -B & -G & 0 & 0 \\ -F & -G & -H & 0 & 0 \\ 0 & 0 & 0 & -Z^2(C+\omega'C^2) & \omega'C^2 Z \\ 0 & 0 & 0 & \omega'C^2 Z & (C-\omega'C^2) \end{bmatrix} \quad (3.4)$$

From (3.3) and (3.4), we can easily obtain the expressions of non-vanishing components of $R_{[ij]}$ such that

$$R_{[14]} = -R_{[41]} = -ZR_{[15]} = ZR_{[51]} = -Z\sigma_1/z\alpha,$$

$$R_{[24]} = -R_{[42]} = -ZR_{[25]} = ZR_{[52]} = -Z\rho_1/z\alpha,$$

$$R_{[34]} = -R_{[43]} = -ZR_{[35]} = ZR_{[53]} = -Z\eta_1/z\alpha. \quad (3.5)$$

It has been observed that $R_{[ij]} R^{[ij]}$ vanishes identically. Therefore expression for Maxwellian $M_{ij}(R_{[\ ]})$ of the tensor $R_{[ij]}$ reduces to

$$M_{ij}(R_{[\ ]}) = -R_{[il]}R_{[km]}g^{lm} = -(1/2)(R_{[il]}R_{[jm]} + R_{[li]}R_{[mj]}). \quad (3.6)$$

From (3.6) the non vanishing components of Maxwellian are calculated as

$$M_{44} = -ZM_{45} = -ZM_{45} = Z^2 M_{55} = -\frac{Z^2}{z^2\alpha^2 m}\{[\sigma_1^2(BH-G^2)+\rho_1^2(AH-F^2)+\eta_1^2(AB-D^2)]$$

$$+ 2[\sigma_1\rho_1(FG-DH)+\sigma_1\eta_1(DG-BF)+\rho_1\eta_1(DF-AG)]\} \quad (3.7)$$

4. Solutions of the field equations (1.1), (1.2) and (1.3) in V_5

In this section we solve the field equations (1.1), (1.2) and (1.3) of Israel and Trollope in higher five dimensional space-time of Ambatkar (2002). The field equation (1.1) is equivalent to

$$S^*_{ij;\mu} = 0, \quad (4.1)$$

which is nothing but the first equation of Einstein's non-symmetric unified field theory and has already been solved by Warade (2006) in [2]. The equation (4.1) enables us to calculate the components of Γ^{*i}_{jk}. Therefore, we can consider the components of Γ^{*i}_{jk} are the same as that of Γ^i_{jk} given in [2] such that

$$\Gamma^k_{11} = [0,0,0,\overline{A}/2CzZ, \overline{A}/2Cz], \quad \Gamma^k_{22} = [0,0,0,\overline{B}/2CzZ, \overline{B}/2Cz],$$

$$\Gamma^k_{33} = [0,0,0,\overline{H}/2CzZ, \overline{H}/2Cz],$$

$$\Gamma^k_{44} = [0,0,0,\frac{1}{z}(-1-\frac{E}{C}-\frac{Z\overline{C}}{2C}+\alpha'), \frac{Z}{z}(1-\frac{2E}{C}+\frac{Z\overline{C}}{2C}+\alpha')],$$

$$\Gamma^k_{55} = [0,0,0,\frac{1}{zZ^2}(\frac{E}{C}-\frac{Z\overline{C}}{2C}+\alpha'), \frac{1}{zZ}(\frac{Z\overline{C}}{2C}+\alpha')],$$

$$\Gamma^k_{12} = \Gamma^k_{21} = [0,0,0,\overline{D}/2CzZ, \overline{D}/2Cz], \quad \Gamma^k_{13} = \Gamma^k_{31} = [0,0,0,\overline{F}/2CzZ, \overline{F}/2Cz],$$

$$\Gamma^k_{14} = -Z\Gamma^k_{15} = [-\frac{ZA'}{z},-\frac{ZB'}{z},-\frac{ZC'}{z},-\frac{\alpha_1}{Z^2C},-\frac{\alpha_1}{ZC}],$$

$$\Gamma^k_{41} = -Z\Gamma^k_{51} = [-\frac{ZA'}{z},-\frac{ZB'}{z},-\frac{ZC'}{z},\frac{\alpha_1}{Z^2C},\frac{\alpha_1}{ZC}],$$

$$\Gamma^k_{23} = \Gamma^k_{32} = [0,0,0,\frac{\overline{G}}{2CzZ},\frac{\overline{G}}{2Cz}],$$

$$\Gamma^k_{24} = -Z\Gamma^k_{25} = [-\frac{ZD'}{z},-\frac{ZE'}{z},-\frac{ZF'}{z},-\frac{\beta_1}{Z^2C},-\frac{\beta_1}{ZC}]$$

$$\Gamma^k_{42} = -Z\Gamma^k_{52} = [-\frac{ZD'}{z},-\frac{ZE'}{z},-\frac{ZF'}{z},\frac{\beta_1}{Z^2C},\frac{\beta_1}{ZC}],$$

$$\Gamma^k_{34} = -Z\Gamma^k_{35} = [-\frac{ZG'}{z},-\frac{ZH'}{z},-\frac{ZI'}{z},-\frac{\gamma_1}{Z^2C},-\frac{\gamma_1}{ZC}],$$

$$\Gamma^k_{43} = -Z\Gamma^k_{53} = [-\frac{ZG'}{z},-\frac{ZH'}{z},-\frac{ZI'}{z},\frac{\gamma_1}{Z^2C},\frac{\gamma_1}{ZC}]$$

$$\Gamma^k_{45} = \Gamma^k_{54} = [0,0,0,\frac{1}{zZ}(1+\frac{Z\overline{C}}{2C}-\alpha'),\frac{1}{z}(\frac{E}{C}-\frac{Z\overline{C}}{2C}-\alpha')] \tag{4.2}$$

where $\quad \alpha' = \dfrac{Z\overline{E}}{2C} - \dfrac{ZE\overline{C}}{C^2} + \dfrac{E^2}{C^2},$

$$A' = [-(BH-G^2)\overline{A} + (DH-FG)\overline{D} - (DG-BF)\overline{F}]/2m,$$

$$B' = [(DH - FG)\overline{A} - (AH - F^2)\overline{D} + (AG - DF)\overline{F}]/2m,$$

$$C' = [-(DG - BF)\overline{A} + (AG - DF)\overline{D} - (AB - D^2)\overline{F}]/2m,$$

$$D' = [-(BH - G^2)\overline{D} + (DH - FG)\overline{B} - (DG - BF)\overline{G}]/2m,$$

$$E' = [(DH - FG)\overline{D} - (AH - F^2)\overline{B} + (AG - DF)\overline{G}]/2m,$$

$$F' = [-(DG - BF)\overline{D} + (AG - DF)\overline{B} - (AB - D^2)\overline{G}]/2m,$$

$$G' = [-(BH - G^2)\overline{F} + (DH - FG)\overline{G} - (DG - BF)\overline{H}]/2m,$$

$$H' = [(DH - FG)\overline{F} - (AH - F^2)\overline{G} + (AG - DF)\overline{H}]/2m,$$

$$I' = [-(DG - BF)\overline{F} + (AG - DF)\overline{G} - (AB - D^2)\overline{H}]/2m,$$

$$\alpha_1 = \frac{1}{z^2}[(-\overline{\sigma}_1 + \frac{\overline{C}}{C}\sigma_1)Z + (A'\sigma_1 + B'\rho_1 + C'\eta_1)Z + (1 - \frac{E}{C})\sigma_1],$$

$$\beta_1 = \frac{1}{z^2}[(-\overline{\rho}_1 + \frac{\overline{C}}{C}\rho_1)Z + (D'\sigma_1 + E'\rho_1 + F'\eta_1)Z + (1 - \frac{E}{C})\rho_1],$$

$$\gamma_1 = \frac{1}{z^2}[(-\overline{\eta}_1 + \frac{\overline{C}}{C}\eta_1)Z + (G'\sigma_1 + H'\rho_1 + I'\eta_1)Z + (1 - \frac{E}{C})\eta_1]. \qquad (4.3)$$

The field equations (1.2) and (1.3) of Israel and Trollope's unified field theory are satisfied identically for the values of (4.2) and the values of $\theta^{[ij]}$ respectively.

In the following section we solve the field equation (1.4) and (1.5) of Israel and Trollope in five dimensional space-time of Ambatkar (2002)

5. Solutions of the field equations (1.4) and (1.5) in V_5

Using the components of Γ^{*i}_{jk} from (4.2) the non-vanishing symmetric components of R^*_{ij} are obtained as follows :

$$-R^*_{44}/Z^2 = R^*_{45}/Z = R^*_{54}/Z = -R^*_{55} = I/z^2, \text{ other } R^*_{ij} = 0 \tag{5.1}$$

where $I = \{\dfrac{\overline{\overline{m}}}{2m} - \dfrac{\overline{m}^2}{2m^2} - \dfrac{\overline{m}\,\overline{C}}{2m\,C} + \dfrac{\overline{m}\,\overline{E}}{2m\,CZ} + [A'^2 + E'^2 + I'^2 + 2(B'D' + G'C' + H'F')]\}$ (5.2)

Substituting the components of M_{ij} and $R_{(ij)} = R^*_{(ij)}$ from (3.7) and (5.1) respectively in field equation (1.4), we obtain the following equation

$$I = \frac{1}{m\alpha}\{[\sigma_1^2(BH - G^2) + \rho_1^2(AH - F^2) + \eta_1^2(AB - D^2)] + 2[\sigma_1\rho_1(FG - DH)$$

$$+ \sigma_1\eta_1(DG - BF) + \rho_1\eta_1(DF - AG)]\}. \tag{5.3}$$

The equation (5.3) is further satisfied under the condition

$$I - \frac{1}{m\alpha}\{[\sigma_1^2(BH - G^2) + \rho_1^2(AH - F^2) + \eta_1^2(AB - D^2)] + 2[\sigma_1\rho_1(FG - DH)$$

$$+ \sigma_1\eta_1(DG - BF) + \rho_1\eta_1(DF - AG)]\} = 0. \tag{5.4}$$

The equation (1.5) of Israel and Trollope's unified field theory is identically satisfied.

We observed that g_{ij} given by (3.4) are the plane wave solutions of the set [I] of field equations of Israel and Trollope's unified field theory in the five dimensional space-time of Ambatkar (2002) provided (5.4) hold.

The following section is dealt with the study of plane wave solutions of the second set of field equations of Israel and Trollope unified field theory in five dimensional space-time.

6. Solution of the set (II) of field equations in V_5

With the choice $bR = 1$, in the second set of field equation, we observed that the components of θ^{ij}, g^{ij}, g_{ij}, $R^{[ij]}$, $R_{[ij]}$ are the same as in the case of field equations of set [I]. These components have already been found in the previous section. The field equations (1.6), (1.7) and (1.8) are the same as first three field equations in the set [I] and have already been solved in the earlier sections.

Substituting the values of $R^*_{(ij)}$, g_{ij} and M_{ij} from (5.1), (3.4) and (3.7) respectively in the field equation (1.10) for $[ij = 11, 12, 13, 22, 23, 33]$, we find that $A = 0$, $D = 0$, $F = 0$, $B = 0$,

$G = 0$ and $H = 0$ which is the contradiction to the fact made in the assumptions of the metric of Ambatkar (2002).

Therefore we have, in the second set of field equations of Israel and Trollope's unified field theory, five dimensional plane wave solutions in space-time of Ambatkar (2002) do not exist.

8. Conclusions

In this paper we have carried out the five dimensional work of Warade (2006) regarding plane wave solutions of the field equations of Einstein's non-symmetric unified field theory to the case of Israel and Trollope's unified field theory and obtained (t/z)-type plane wave solutions in higher five dimensional space-time of Ambatkar (2002).

Thus our work regarding plane wave solutions is in the frameworks of five dimensional space-time which is generalized to that of Takeno (1961).

We observed that the field equation (1.1) of Israel and Trollope is equivalent to the first equation of Einstein's non-symmetric unified field theory and has already been solved by Warade (2006). The field equation (1.2) of Israel and Trollope's unified field theory is satisfied identically for the values given in (4.2). The field equation (1.3) of Israel and Trollope's unified field theory is also identically satisfied by substituting the values of $\theta^{[ij]}$ and the equation (1.5) of Israel and Trollope is identically satisfied.

We think that these (t/z) - type higher five dimensional plane wave solutions should bring some additional information and therefore, they need to be further investigated.

It is to be noted that in the second set of field equations of Israel and Trollope's unified field theory, this type of five dimensional plane wave solutions in the space-time of Ambatkar (2002) do not exist.

References

[1] J. K Jumale (2006), Some plane wave solutions of field equations in general relativity. *Ph.D. Thesis, R T M Nagpur University, Nagpur.*

[2] D.N. Warade (2006), Some plane wave solutions of field equations and their geometric properties in general relativity. Ph.D. thesis, Nagpur University, Nagpur

[3] Urkude et. al (2011), Five dimensional plane wave solutions in Israel and Trollope's unified field theory. [*Communicated*]

[4] Ambatkar B G (2002), Some geometrical and physical aspects of H and F_{ij}". Ph.D. *Thesis, Nagpur University, Nagpur.*

[5] W. Israel (1961), New possibilities for a unified field theory, *J. Math. Phy.*, R. and Trollope 777-786.

[6] H. Takeno (1961), Some plane wave like solutions of Non-symmetric unified field theory. *Tensor N. S.* 11, 263-268, *Japan*

[7] Takeno H (1961), The mathematical theory of plane gravitational wave in general relativity. *Scientific reports of the research institute for theoretical physics Hiroshima University Hiroshima-Ken Japan.*

[8] Kaluza T (1921), *Sitz preuss. Akad. Wiss.* **D 33**, 966

[9] Kelin O (1926), Z. *phys.* 895

[10] Wesson P S (1983), *Astro. Astrophys.*, **119**, 145

[11] Wesson P S (1984), *Gen. Rel. Grav.* **16**, 193

[12] Reddy D.R.K. (1999), *Astrophy.Space.Sci.*, 1-5

[13] K.B.Lal, The wave solutions of the field equations of Einstein's and N. Ali (1970), Bonnor's and Schrodinger's non-symmetric unified field Theories in a generalized Takeno space-time,Tensor (N.S.), 21, 243-249

[14] R. P. Srivastava (1974),Wave solutions of unified field theories in generalized Takeno space-times. Ph.D.Thesis, Gorkhapur University, Gorkhapur.

[15] A. Pradhan (1977), Wave solutions of field equations in general relativity and non-symmetric unified field theories in a generalized Peres space-time. Ph.D. thesis, Gorkhapur University, Gorkhapur.

[16] Thengane et.al (2000), Plane wave solutions of field equations $R_{ij} = 0$ in *N*-dimensional Space-time. *PP 5-8, Post Raag Reports No.351, 2000, Japan.*

[17] L. S. Ladke (2004), Some aspects of the plane wave solutions in general relativity. Ph.D. thesis, Nagpur University, Nagpur

Essay

If the LHC Particle Is Real, What Is One of the Other Possibilities than the Higgs Boson?

Huping Hu[*] & Maoxin Wu

ABSTRACT

In the prespacetime model, an unspinized particle governed by a matrix law is the precursor of all spinized particles and thus steps into the shoes played by the Higgs particle. We speculate here that what has been found at the LHC, if real, is plausibly the unspinized particle of the prespacetime model. The wave function of a fermion or boson is respectively a bispinor or bi-vector but that of the unspinized particle is two-component complex scalar field. Thus, it may have different behavior than that of either the boson or fermion which may be detectable at LHC.

Key Words: LHC, Higgs boson, unspinized particle, Prespacetime Model.

Hints of a New LHC Particle

Based on the CERN announcement on December 13, 2011, the hints of a new particle commonly attributed to Higgs boson of the Standard Model have been observed at the Large Hadron Collider. Philip E. Gibbs has done some important combinations and calculations and concluded that "I get an overall probability for such a strong signal if there is no Higgs to be about 1 in 30…everything considered I take the observed result to be a two sigma effect. [1-2]."

Prespacetime Model in a Nutshell

We have previously formulated a prespacetime model of elementary particles and four forces [3]. The model illustrates how self-referential hierarchical spin structure of the prespacetime provides a foundation for creating, sustaining and causing evolution of elementary particles through matrixing processes embedded in prespacetime:

$$1 = e^{i0} = 1e^{i0} = Le^{-iM+iM} = \frac{E^2 - m^2}{\mathbf{p}^2} e^{-ip^\mu x_\mu + ip^\mu x_\mu} =$$

$$\left(\frac{E-m}{-|\mathbf{p}|}\right)\left(\frac{-|\mathbf{p}|}{E+m}\right)^{-1}\left(e^{-ip^\mu x_\mu}\right)\left(e^{-ip^\mu x_\mu}\right)^{-1} \rightarrow \quad (1)$$

$$\frac{E-m}{-|\mathbf{p}|} e^{-ip^\mu x_\mu} = \frac{-|\mathbf{p}|}{E+m} e^{-ip^\mu x_\mu} \rightarrow \frac{E-m}{-|\mathbf{p}|} e^{-ip^\mu x_\mu} - \frac{-|\mathbf{p}|}{E+m} e^{-ip^\mu x_\mu} = 0$$

$$\rightarrow \begin{pmatrix} E-m & -|\mathbf{p}| \\ -|\mathbf{p}| & E+m \end{pmatrix} \begin{pmatrix} a_{e,+}e^{-ip^\mu x_\mu} \\ a_{i,-}e^{-ip^\mu x_\mu} \end{pmatrix} == L_M \begin{pmatrix} \psi_{e,+} \\ \psi_{i,-} \end{pmatrix} = L_M \psi = 0 \qquad (2)$$

$$\rightarrow \begin{pmatrix} E-m & -\boldsymbol{\sigma}\cdot\mathbf{p} \\ -\boldsymbol{\sigma}\cdot\mathbf{p} & E+m \end{pmatrix} \begin{pmatrix} A_{e,+}e^{-ip^\mu x_\mu} \\ A_{i,-}e^{-ip^\mu x_\mu} \end{pmatrix} = L_M \begin{pmatrix} \psi_{e,+} \\ \psi_{i,-} \end{pmatrix} = L_M \psi = 0 \qquad (3)$$

or

$$\rightarrow \begin{pmatrix} E-m & -\mathbf{s}\cdot\mathbf{p} \\ -\mathbf{s}\cdot\mathbf{p} & E+m \end{pmatrix} \begin{pmatrix} A_{e,+}e^{-ip^\mu x_\mu} \\ A_{i,-}e^{-ip^\mu x_\mu} \end{pmatrix} = L_M \begin{pmatrix} \mathbf{E} \\ i\mathbf{B} \end{pmatrix} = L_M \psi = 0 \qquad (4)$$

In the above, Equation (2) governs unspinized particles, Equation (3) governs spin-1/2 particles after spinization from (2); and Equation (4) governs spin-1 particles after spinization from (2).

Implication of LHC Finding (If Real) for Prespacetime Model

Traditionally, a spinless particle is presumed to be described by the Klein-Gordon equation and is classified as a boson. However, we have suggested in [3] that Kein-Gordon equation is a determinant view of a fermion, boson or an unspinized particle and the latter is neither a boson nor a fermion but may be classified as a **third state of matter** described by the unspinized equation (2) above in Dirac form. The Weyl (chiral) form is given below:

$$\begin{pmatrix} E-|\mathbf{p}| & -m \\ -m & E+|\mathbf{p}| \end{pmatrix} \begin{pmatrix} a_{e,l}e^{-ip^\mu x_\mu} \\ a_{i,r}e^{-ip^\mu x_\mu} \end{pmatrix} = L_M \begin{pmatrix} \psi_{e,l} \\ \psi_{i,r} \end{pmatrix} = L_M \psi = 0 \qquad (5)$$

The wave function of a fermion or boson is respectively a bispinor or bi-vector but that of the third state is a two-component complex scalar field. In the prespacetime model, the third state of matter is the precursor of both fermionic and bosonic matters/fields before fermionic or bosonic spinization.

Thus, it steps into the shoes played by the Higgs particle in the Standard Model and may be what has been seen at the LHC, if it is real. The third state of matter may have different behavior from that of either the boson or fermion which may be detectable at LHC.

References

1. Gibbs, P. E. (2011), Higgs Boson Live Blog: Analysis of the CERN Announcement. Prespacetime Journal, 2(13): pp. 2021-2043; http://blog.vixra.org/2011/12/13/the-higgs-boson-live-from-cern/

2. Gibbs, P. E. (2011), Has CERN Found the God Particle? A Calculation, Prespacetime Journal, 2(13): pp. 2044-2052; http://blog.vixra.org/2011/12/16/has-cern-found-the-god-particle-a-calculation/

3. Hu, H. & Wu, M. (2010), Prespacetime Model of Elementary Particles, Four Forces & Consciousness. Prespacetime Journal, 1(1): pp. 77-146.

Essay

What is Reality in a Holographic World?

James Kowall[*]

Abstract

The nature of a holographic world is described. This scientific description of the world is based upon the assumptions of modern theoretical physics. These natural assumptions are inherent in any unified theory, such as string theory, and in any theory of the creation of the world, such as inflationary cosmology. At their most basic level, these are the assumptions of the equivalence, uncertainty and action principles, along with the second law of thermodynamics. Any world consistent with these fundamental principles is easily shown to be a holographic world. The mathematical consistency of such a holographic world also implies something about the nature of consciousness. If that mathematical consistency is followed to its logical conclusion, in the sense of the Gödel incompleteness theorems, this scientific description of the world also has something to tell us about the nature of reality. What this scientific description of the world tells us about the nature of reality is compared to what mystics have told us about reality throughout human history.

A holographic description of the world is an observer-centric description that satisfies the covariant entropy bound of the holographic principle. In such a world, consensual reality is not a single objective reality, but many entangled worlds that share information with each other, each defined on its own viewing screen, and each observed from its own point of view. The self-concept is understood in terms of the encoding of information on the viewing screen, and the expression of personal will and universal will is understood in the sense of the flow of energy. The nature of the Self is understood as a presence of consciousness that arises at a focal point of perception, while the perceivable world arises on a viewing screen. In a nondual sense, the Source of any such world, and the Source of any individual consciousness, is understood as a void of undifferentiated consciousness.

Key words: Reality, holographic world, consciousness.

The eternal mystery of the world is its comprehensibility.

Everyone who is seriously involved in the pursuit of science becomes convinced that a spirit is manifest in the laws of the Universe.

Albert Einstein

What is reality? There is no immediately obvious way to answer this question. Any answer that we can give depends upon our assumptions. There is no way to avoid the fact

---

[*] Correspondence: James Kowall, PhD, MD. E-mail: jkowall@earthlink.net

that any answer to this question is assumption dependent. There is however a natural way to approach this question, which is the scientific method.

The scientific method is fundamentally based upon observation. We observe the world, and we deduce properties of the world from our observations. From our observations of the world we then construct theories about the world. Those theories allow us to make predictions, which we can test in subsequent observations of the world. It is often stated that the scientific method is based upon experimental observation, but that is not quite correct. An experimental set-up is always a part of the world we perceive. The experimenter is as much a part of the world we perceive as is the experimental apparatus.

What about the nature of consciousness? Is the nature of consciousness a part of the same world we perceive with our observations of the world? If we assume that consciousness arises within the same world with the things perceived within that world, that assumption is a paradox of self-reference, and leads to logical inconsistency, since it implicitly identifies consciousness with something that consciousness perceives within that world.

How can the nature of consciousness be identical to something that consciousness perceives within the world? Science has no answer to this question, since the scientific method is based on observation, and assumes the existence of consciousness.

Robert Ellis has written an article entitled "*Taking the 'Meta' out of Physics*", which is the focus of attention of this focus issue. Early in this article he states that "quantum physics cannot give us metaphysical information, and that metaphysical claims supported by quantum physics are at best an irrelevant distraction from the Buddha's key insights".

What exactly are those key insights of the Buddha? Who is the one that is having those insights? Who is the knower of that knowledge? It is just not possible to answer these questions unless we understand the nature of consciousness, which of course is our own insight into this question concerning the Buddha's insights. In a similar way, our insights into the nature of quantum physics also depend on the nature of our consciousness.

Ellis admits he is unaware of the full implications of modern physics. In a dismissive way he states "as a philosopher, I do consider myself qualified to comment on the general conditions surrounding knowledge claims. It seems that quantum physicists have become gods, if they really claim to be able to support metaphysical beliefs from finite scientific observation". But the question is still the same: what is the true nature of the knower?

Simply stated, it is impossible to take the 'meta' out of physics since it is impossible to take the observer out of physics. It is impossible to take the knower out of knowledge. All metaphysical discussions are inherently about the nature of the observer and the knower. There is no physical theory of the observer because consciousness cannot be explained physically. Everything our physical theories of the observable world describe is some physical thing observed by an observer. The observer is inherent in our most basic scientific principles, like the principle of equivalence. All the scientific debate about the correct interpretation of quantum theory is about the nature of observation. Both physics and metaphysics place the observer at the center of this discussion.

If Ellis wants to take the 'meta' out of physics he either has to take the observer out of physics (which is impossible) or he has to give a physical explanation of the nature of consciousness (which is equally impossible). Until he does one or the other, what he has to say about the Buddha's insights, or anyone else's insights, does not make any sense.

Stephen Hawking addresses this conundrum about the nature of observation and quantum theory (and all physical theories) with the following two statements:

"I don't demand that a theory correspond to reality because I don't know what it is. Reality is not a quality you can test with litmus paper. All I'm concerned with is that the theory should predict the results of measurements" (Penrose 2005, 29.1).

"Personally, I get uneasy when people, especially theoretical physicists, talk about consciousness. Consciousness is not a quality that one can measure from the outside" (Penrose 1999, 171).

We may find this state of affairs uneasy, but the fact remains it is impossible to discuss any physical theory of any physical world without mentioning the observer of that world.

When we refer to the observer of our own physical world, we are referring to our own consciousness. This is the key insight that Ellis does not seem to realize. It is not even possible to discuss the nature of a physical world without the observer of that world.

There is a modern physical theory of the physical world that almost seems to demand of us that we discuss the nature of the observer of that physical world. This kind of physical description of the physical world is inherent in all modern unified theories, which unify the fundamental forces of nature. This kind of physical description of the physical world is also inherent in all modern theories of the creation of the physical world.

Our modern physical theories of the physical world incorporate a fundamental principle, referred to as the holographic principle of quantum gravity (Susskind 1994, 1). This is a very strange principle precisely because we cannot understand what it has to tell us about the physical world unless we examine what it has to tell us about the observer of that world. Simply stated, without the observer of that world, there is no physical world.

Susskind tells us that he is taking us into a "very strange territory" with the holographic principle (Susskind 2008, 298-299). He is some of what he has to say about it:

"...the three-dimensional world of ordinary experience – the universe filled with galaxies, stars, planets, houses, boulders, and people – is a hologram, an image of reality coded on a distant two-dimensional surface. This new law of physics, known as the Holographic Principle, asserts that everything inside a region of space can be described by bits of information restricted to the boundary. To put it in concrete terms, consider the room I am working in. I in my chair, the computer in front of me, my messy desk piled high with papers I'm afraid to throw out – all that information – is precisely coded in Planckian bits, far too small to see but densely covering the wall of the room. Or instead, think of everything within a million light-years of the Sun. That region also has a boundary – not physical walls, but an imaginary mathematical shell – that contains everything within it: intersteller gas, stars, planets, people, and all the rest. As before, everything inside that giant shell is an image of microscopic bits spread over the shell. Moreover, the required number of bits is at most one per Planck area. It is as if the boundary – office walls or mathematical shell – were made of tiny pixels, each occupying one square Planck length, and everything taking place in the interior of the region is a holographic image of the pixilated boundary."

Susskind describes how all the information for the images of a physical world that appear within space is encoded in terms of bits on information on the boundary surface of that space. In this sense, that bounding surface acts just like a holographic viewing screen that projects perceivable images to a focal point of perception (Bousso 2002, 28). If that bounding surface is a sphere, then that focal point of perception is at the central point of view of that sphere. That physical world of images demands of us that we inquire into the nature of the consciousness of the observer present at that focal point of perception.

The scientific method is fundamentally about the things we observe in the world. In the language of modern theoretical physics, those things can always be deconstructed into the nature of information and energy. The scientific method is reductionistic in nature, and reduces all things in the world down to the fundamental nature of information and energy. The remarkable thing about modern theoretical physics is that it ultimately reduces all information and energy down to its fundamental holographic nature.

This holographic description of the world is fundamentally expressed as the holographic principle of quantum gravity (Susskind 2008, 290). Any scientific description of the world that incorporates the equivalence, uncertainty and action principles, along with the second law of thermodynamics, can easily be shown to be a holographic world.

The holographic nature of the world describes at the most fundamental level possible how all information and energy is encoded in the world. But what does that fundamental description of the world tell us about the fundamental nature of consciousness? What is the nature of the consciousness that perceives that holographic world?

There is a straightforward answer that can be given to this question if we have the fortitude to follow the scientific method to its logical conclusion. The answer is not an easy one to accept, but it is the only answer possible if we require our scientific description of the world to possess the quality of logical consistency.

There is something fundamentally wrong with the conventional scientific concept of the world held in the minds of most scientists. That concept is the idea that the world consists of matter and energy that exist within space and time. The usual idea of matter and energy is the atomic hypothesis, which says that at a fundamental level all matter and energy is composed of point particles, like the electron and photon. Point particles exist at points of space, and trace out paths through space over the course of time. Quantum theory only extends the classical idea of a point particle to a sum over all possible paths.

But those point particles only exist if there is a pre-existing space and time for particles to exist within. This is the kind of scientific paradigm described by any quantum field theory. The usual idea of that pre-existing space and time is flat Minkowski space-time geometry (Zee 2003, xv). A quantum field $\Psi(x,t)$ is interpreted as a probability amplitude that specifies the probability that the particle excitation of energy associated with that field can be measured at position x and time t in that pre-existing space-time geometry. The probability for the particle to propagate between two space-time points is expressed in terms of a sum over all possible paths that connect those two points (Zee 2003, 9).

Quantum field theory assumes the existence of a vacuum state from which all particle excitations of field energy arise (Zee 2003, 19). The nature of the vacuum state is conceptualized as empty space, and is the ground state from which all excited states arise. A particle is an excitation of field energy. The nature of a force is conceptualized as an exchange of particles between other particles (Zee 2003, 27). Quantum field theory is more complicated than ordinary quantum mechanics due to virtual particle-antiparticle pairs that can arise within the path of any point particle (Zee 2003, 55).

The path of any particle is drawn as a diagram that connects two space-time points (Zee 2003, 41). In terms of diagrams, the virtual particle-antiparticle pairs are drawn as closed loops. Closed loops can arise even within the vacuum state, and are interpreted as virtual particle-antiparticle pairs that are created out of nothing and annihilate back into nothing within a short period of time due to quantum uncertainty in energy (Zee 2003, 57).

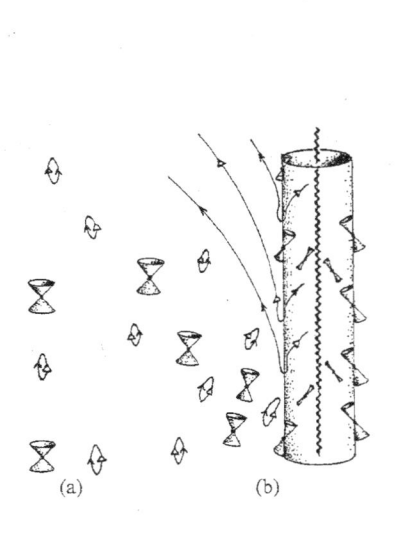

Fig. 30.11 Hawking's 'intuitive' derivation of Hawking radiation. (a) Far from the hole, virtual particle-anti-particle pairs are continually produced out of vacuum, but then annihilated in a very short time (see Fig. 26.9a). (b) Very close to the hole's horizon, we can envisage one of the pair falling into the hole, the other escaping to external infinity. For this, the virtual particles both become real, and energy conservation demands that the ingoing particles have negative energy. This it can do, because the Killing vector κ becomes spacelike inside the horizon. (If κ^a is spacelike, the conserved energy $p_a \kappa^a$ can be negative, where p_a is the particle's 4-momentum.)

The vacuum state is a state of zero energy, but due to quantum uncertainty, the vacuum state has quantum fluctuations in energy. In the sense of energy conservation, the virtual antiparticle carries an equal but opposite amount of energy as the virtual particle, so that the total energy of a vacuum fluctuation adds up to zero (Penrose 2005, figure 30.11).

Relativity theory is also a field theory, since it describes the gravitational field, but it is a most unusual field theory, since the gravitational field describes the dynamical nature of space-time geometry. In relativity theory space-time geometry is dynamical, and there is no pre-existing space and time for point particles to exist within. The gravitational field and the curvature of space-time geometry are represented in terms of a metric (Penrose 2005, 17.9), which describes the amount of proper time that passes on any path of a point particle that connects two space-time points (Penrose 2005, figure 17.15). The concept of proper time is analogous to length along the path in a curved space-time geometry.

The problem with the conventional scientific paradigm of quantum field theory is that it contradicts relativity theory, which describes the dynamical nature of space-time geometry. Relativity theory describes the gravitational field. That field describes the dynamical nature of space-time geometry, but it cannot be quantized. According to relativity theory, there is no such thing as a pre-existing space and time for point particles to exist within. If there was such a thing, then relativity theory could be quantized, and would result in the point particle we call the graviton. The graviton would exist at a point in that pre-existing space and time, which according to relativity theory doesn't exist. This is the ultimate chicken and egg problem. There is just no way to quantize relativity theory as a field theory (Zee 2003, 434; Susskind 2008, 331).

The second problem with this conventional scientific paradigm is the problem of consciousness. In a strange way, the problem of consciousness is related to both relativity theory and quantum theory. Relativity theory is based upon the principle of equivalence, which expresses the equivalence of all observational points of view. Quantum theory says that everything that is observable in the world is specified by an observable value of the quantum state, and is observed by an observer. Relativity theory expresses the equivalence of all observers, present at all points of view. The problem with the conventional paradigm of point particles that exist within some pre-existing space and time is its logical inconsistency, which results in paradoxes of self-reference. There is a logical contradiction if consciousness somehow arises within that pre-existing space and time from the behavior of the point particles that exist within that space and time, since consciousness is what observes the behavior of those point particles. This logical contradiction is the idea that the observer somehow arises from the behavior of some observable thing that it can observe, which is a paradox of self-reference, since it equates the observer with an observable value (Goldstein 2005, 165; Penrose 1999, 112).

The Gödel incompleteness theorems prove that the only way any science, based on the logical consistency of mathematics, is free of these paradoxes of self-reference, is if the observer is 'outside' of whatever observable values it observes, just like the viewer of a computer viewing screen is always 'outside' of the computational information displayed on that viewing screen. A logical contradiction arises if consciousness somehow arises in the same world that behavior arises within, since consciousness is what observes that behavior (Penrose 2005, 34.6). The second incompleteness theorem proves any consistent mathematical system as complex as counting natural numbers can never prove its own consistency. The 'proof of consistency' is always 'outside'. (Goldstein 2005, 183). Consciousness is always 'outside', since it is what 'knows' about that logical consistency.

The holographic principle explains how this is possible. The consciousness of the observer that views the viewing screen is always present at a point of view that is outside the viewing screen. That presence of consciousness does not arise in the same world that behavior arises within. That behavior arises on a viewing screen from the way information is encoded on the viewing screen, and the way energy flows through that world over a sequence of events, as viewing screens are animated like the frames of a movie (Penrose 2005, figure 17.1). As energy flows, information is coherently organized into animated forms of information. Those forms appear three dimensional since they are holographic. Animated forms of information are displayed on the viewing screen over a sequence of events, and the forms tend to replicate in form due to coherent organization. Coherent organization is what allows for self-replication of form over a sequence of events displayed on viewing screens. Forms coherently organized on the viewing screen are projected like images to a focal point of perception (Penrose 2005, figure 15.13). The consciousness of the observer is always outside the screen, present at a point of view.

What screen? What observer? What exactly are we talking about? Where exactly is the screen? Where exactly is the observer? There is no way to answer these questions without the principle of equivalence. Relativity theory is fundamentally based on the principles of relativity and equivalence. The principle of relativity expresses the constancy of the speed of light as observed from all points of view. The principle of equivalence expresses that every force is equivalent to an accelerating frame of reference. We can always place an observer at the origin of any frame of reference. An accelerating frame of reference always has an event horizon, which is a two dimensional surface that is as far as the observer can see things in space due to the constancy of the speed of light. The surface of the event horizon is the viewing screen. This is usually represented in terms of a Penrose diagram (Penrose 2005, 27.12), which describes the nature of the 'observable world' as observed from the central point of view of that accelerating frame of reference.

The key insight of the holographic principle is that an accelerating frame of reference, with an observer present at the central point of view, can arise even within empty space. The observer's entire observable world is limited by the event horizon. As the observer arises, an event horizon also arises, which is a far as the observer can see things in space due to the constancy of the speed of light (Penrose 2005, figure 27.16). Where does the point of view of the observer arise? Where does the two dimensional surface of the event horizon arise? They both arise in empty space.

Forces are inherently geometrical in nature since they are equivalent to accelerations, as observed from the point of view of an accelerating frame of reference. That accelerating frame of reference always has an event horizon, which is a two dimensional surface that is as far as the observer present at that central point of view can see things in space due to the constancy of the speed of light. The principle of relativity expresses the constancy of the speed of light observed by all observers present at all points of view in empty space. The equivalence of any force with an accelerating frame of reference expresses the equivalence of all points of view in empty space (Greene 1999, 61).

The equivalence of all points of view in empty space is explicitly demonstrated in relativity theory with general focusing and projection theorems (Bousso 2002, 26, 36), which prove the number of fundamental degrees of freedom in any region of space are defined upon a bounding surface of space. The information content for those degrees of freedom is measured by the area of the bounding surface, and can always be projected to a central point of view, which is a focal point of perception. The bounding surface is an event horizon, which is as far as the observer present at that central point of view can see things in space. Every accelerating frame of reference has an event horizon, which is as far as the observer can see things. This relationship was discovered when the entropy of black holes was first calculated. Entropy measures disordered information, which is inherently related to disordered kinetic energy. If too many degrees of freedom are excited in some spherical region of space, the region becomes very massive and must gravitationally collapse into a black hole with an event horizon. The entropy of the black hole is proportional to the surface area of the event horizon.

This relationship is easily shown (Susskind 2008, 152). The only assumptions necessary are the equivalence and uncertainty principles, and the second law of thermodynamics. Any world that incorporates these basic scientific principles must be holographic.

Why is the holographic principle implied in such a world? In quantum theory, if we want to observe the behavior of a point particle as it moves through space, we have to use some kind of radiation, such as light, which is electromagnetic radiation. We shine the radiation at the particle, and observe how the radiation scatters off the particle. We can only know about the position and the motion of the point particle because of the scattered radiation we observe. Quantum theory tells us that if we want to look at smaller distance scales we have to use higher frequencies of radiation, since the distance scale we can probe with light is set by the wavelength of the light, λ, which is related to the frequency of the light waves, ν, and the speed of light, as $c=\lambda\nu$. In quantum theory, a higher frequency of vibration corresponds to a higher energy, as $E=h\nu=hc/\lambda$, and so we must use higher energies to look at smaller distances. This is where relativity theory comes in. At some point, we focus so much energy into such a small region of space that we create a black hole. Once the black hole is created, the only radiation we can see scattered off the surface of the black hole is Hawking radiation. If we shine higher energy radiation at the black hole, we only create a bigger black hole, with a larger event horizon, which only radiates away lower frequencies of Hawking radiation back to us (Susskind 2008, 268).

There is an ultimate distance scale that we can probe, which is the Planck length. This is the length that corresponds to a frequency of radiation where enough energy is concentrated into a small enough region of space that a black hole is forced to form. Relativity theory demands that it is impossible to probe smaller distance scales than the Planck length. The reductionistic tendency to probe smaller distance scales with larger energies must finally come to an end at the Planck scale. The way that reductionistic tendency comes to an end is inherently holographic in nature. All the information for the black hole is encoded on the surface of the event horizon. An event horizon is a two dimensional surface that encodes quantized bits of information.

Relativity theory predicts event horizons. The event horizon of a black hole is a two dimensional surface of radius R where the acceleration due to gravity is so strong that even light cannot escape. The radius of the event horizon is given in terms of the mass of the black hole M, as $R=2GM/c^2$, where G is the gravitational constant and c is the speed of light. This result follows from the acceleration due to gravity on the surface of a gravitating body of mass M and radius R, $g=GM/R^2$. The acceleration of gravity is so strong at the event horizon that even light cannot escape. No physical signal that originates from inside a black hole can ever cross the event horizon.

It is instructive to examine this relationship a bit further. The classical equations of motion for an object, like a point particle, are usually expressed as Newton's law, which states that the force F applied to that object is equal to its mass m times its acceleration a, or $F=ma$. The other law Newton discovered is the law of gravity, which states the force of gravity applied to an object of mass m by a gravitating body of mass M, when the two objects are separated by a distance R, is given by $F=GMm/R^2$. Einstein postulated the principle of equivalence based upon the fact that gravitational mass is the same as inertial mass, and so the force of gravity is equivalent to an accelerating frame of reference. The acceleration of gravity due to a gravitating body of mass M at a distance R from that body is independent of the mass m of a particle that the gravitational force acts upon, and is given by $a=g=GM/R^2$. The reason the gravitational field can be conceptualized as the curvature of space-time geometry is precisely due to the fact the force of gravity is always equivalent to an acceleration. An acceleration is geometrical in nature, just as position in space and velocity through space are geometrical in nature.

There is an easy way to determine the radius of the event horizon from classical principles (Susskind 2008, 48). The concept of escape velocity is inherent in both classical mechanics and relativity theory, and represents the amount of kinetic energy needed to overcome the potential energy of gravitational attraction. The amount of gravitational potential energy at the surface of a gravitating body of radius R and mass M experienced by a particle of mass m is given by $PE=-GMm/R$. In classical physics, the kinetic energy of that particle as it moves away from the gravitating body with velocity v is given by $KE=\frac{1}{2}mv^2$. The particle just escapes if it has just enough kinetic energy to overcome the gravitational attraction, which is determined by total energy $E=KE+PE=0$, and gives the escape velocity as $v^2=2GM/R$. For a black hole, we equate escape velocity with the speed of light, $v=c$, and determine the radius of the event horizon as $R=2GM/c^2$.

There is an easy way to see how the event horizon of a black hole encodes quantized bits of information. The energy of a Hawking photon radiated away from the event horizon is given by $E=hc/\lambda$. The amount of gravitational potential energy that photon experiences at the surface of the event horizon is given by $PE=-GmM/R$. The 'effective mass' of the photon is related to its energy by $E=mc^2$. A photon is only bound to a black hole since its potential energy of gravitational attraction outweighs its kinetic energy, just like an electron bound to a proton by the electromagnetic force. The electron can orbit the proton in a circular orbit of radius r. The ground state orbit of a hydrogen atom is specified by an electron wave function with a wavelength $\lambda=2\pi r$, as a single wavelength fits into the circumference of the orbit. This wavelength determines the amount of energy needed to ionize the electron, as it escapes away from the proton (Feynman 1963, I 38-5). In a similar way, a photon that is gravitationally bound to a black hole has a wavelength set by the radius of the event horizon, and has a similar 'ionization energy'.

The smallest quantized bit of information radiated away from the black hole is a photon with wavelength equal to the circumference of the event horizon, or $\lambda=2\pi R$, which gives that photon an energy of $E=hc/2\pi R$. That is the smallest bit of energy the black hole can emit, just like an ionized electron that escapes away from its ground state orbit around a proton in a hydrogen atom. As that photon is emitted, the radius $R=2GM/c^2$ of the event horizon decreases, since energy is radiated away from the black hole, and the mass M of the black hole decreases by an amount $\Delta M=m=E/c^2=h/2\pi Rc$. The emitted photon carries energy away from the black hole, which results in a decrease in the radius of the event horizon as $\Delta R=2G\Delta M/c^2=2Gh/2\pi Rc^3$. The area of the event horizon is $A=4\pi R^2$. The change in area of the event horizon that corresponds to this change in radius is given by $\Delta A/A=2\Delta R/R=8Gh/c^3A$, or $\Delta A=8Gh/c^3=16\pi\ell^2$, where ℓ is the Planck length. The natural definition of a Planck area is $a=4\pi\ell^2$. As a quantized bit of information is radiated away from the black hole, the area of the event horizon decreases by four Planck areas. If we imagine that four Planck areas act like a pixel on a screen that encodes a quantized bit of information, then the total number of bits of information encoded on the event horizon, b, is the surface area divided by the area of a pixel. This is exactly the kind of relationship that Hawking found. The correct relationship is given by $b=A/4\ell^2$.

The above argument is the heuristic explanation of Bekenstein (Susskind 2008, 152). Hawking's argument is mathematically more sophisticated, and uses quantum field theory in a curved space-time geometry. At an intuitive level, Hawking's explanation is based upon the apparent separation of virtual particle-antiparticle pairs at the event horizon, as observed by a distant observer (Susskind 2008, 171; Penrose 2005, figure 30.11). That apparent separation of virtual pairs at the horizon is inherently related to the encoding of quantized bit of information on the horizon, with one bit of information encoded per pixel on the screen. The two dimensional surface of the event horizon acts as a holographic viewing screen that projects the three dimensional images of things observed in space to a focal point of perception. Both the event horizon and the observer arise within empty space. Where is that focal point of perception? In the sense of inflationary cosmology, everything perceived within that space, which includes the black hole, is observed from the central point of view of a cosmic event horizon (Susskind 2008, 304).

What about the usual interpretation of quantum field theory that any quantum field is a probability amplitude that species the probability of measuring some quantized physical property of the point particle at some point in space-time? That probability amplitude corresponds to the projection of a holographic image from the surface of the horizon, which acts as a holographic viewing screen. All the information is defined on the viewing screen, and is perceived at a point of view. The point particle description of physical reality in terms of a quantum field theory is a holographic description. The more fundamental description is the viewing screen description, since that is where all the fundamental bits of information are defined. In this sense, the propagation of a light wave is like the projection of an image from the viewing screen to the central point of view.

The quantum field $\Psi(x,t)$ that describes the propagation of a light wave is a probability amplitude that specified the probability with which a photon can be measured at position x at time t. The quantum field has both wave-like properties and particle-like properties due to the sum over all possible paths of the photon (Susskind 2008, 77). In terms of the viewing screen description, all the information for the photon is coherently encoded on the viewing screen, and the measurement of any photon property in space is like the holographic project of an image from the viewing screen to a point of view. The nature of time arises as images are animated over a sequence of events, like the frames of a movie.

Relativity theory expresses the fundamental nature of consciousness through the principle of equivalence, which expresses the equivalence of observers present at all points of view in empty space. Contrary to what is often assumed in the scientific literature, quantum theory does not express the fundamental nature of consciousness. The fundamental nature of quantum theory is the uncertainty principle, which describes how something is created from nothing, as virtual particle-antiparticle pairs spontaneously arise from the vacuum state. Virtual pairs appear to separate at an event horizon, as observed by the observer present at the central point of view. That separation of matter from antimatter, called Hawking radiation, is the essence of the holographic principle (Penrose 2005, 30.7).

Inflationary cosmology (Penrose 2005, 28.4) is another idea that has something very important to tell us about the nature of consciousness. Inflationary cosmology explains the nature of the big bang event, which is how the universe is created, and is supported by a lot of observational evidence, but it inevitably leads to the conclusion that multiple universes exist. Multiple universes are referred to as an ensemble of universes, which describe all possible ways in which the universe can be created and evolve. An ensemble of universes is understood both in the sense of quantum theory and thermodynamics.

Every possible way in which the universe can be created and evolve is described by a state of information. Quantum theory describes how quantized bits of information are encoded in any state of information, and thermodynamics describes how states of information evolve over time. The holographic principle of quantum gravity explains how information is encoded on the surface of an event horizon, which acts as a holographic viewing screen, with one fundamental bit of quantized information encoded per fundamental pixel on the screen. An ensemble of universes describes all possible ways in which information can be encoded in an initial state and evolve over time.

An ensemble of universes is often described as 'bubbles in the void' (Greene 2011, 66). The surface of a bubble is called a cosmic event horizon (Susskind 2008, 435). In an exponentially expanding universe like ours, there is always a cosmic event horizon, where the universe at that point appears to expand at the speed of light, as observed by the observer present at the central point of view. Since nothing can ever travel faster than the speed of light, that horizon is as far out as that observer can see things in space. The radical nature of what inflationary cosmology tells us about consciousness is that the observer is located at the central point of view of a cosmic event horizon.

The exponential expansion of the universe implies a cosmological constant, which in the sense of quantum field theory arises from vacuum energy (Zee 2003, 434). That universal expansion is caused by 'dark energy', or vacuum energy, which causes the universe to repel itself, and is a kind of anti-gravity. The universe exponentially expands from the big bang event due to that repulsion. Dark energy arises from the vacuum state due to quantum uncertainty. This is usually described as the virtual creation of particle-antiparticle pairs, or a closed-loop process. Virtual particle-antiparticle pairs are created out of nothing, and normally annihilate back into nothing within a short period of time, as specified by the uncertainty principle. In some sense, the virtual antiparticle carries an equal but opposite amount of energy as the virtual particle, so that the total energy of this virtual process adds up to zero (Penrose 2005, 30.7). Virtual pairs appear to separate at a cosmic event horizon, as observed by the observer at the central point of view of that spherical surface, which is how a universe of matter is created.

The cosmic event horizon inflates in size from the big bang since there is an instability in amount of dark energy, due to a phase transition that occurs as the universe expands and cools, similar to super-cooled liquid water that freezes into ice. The big bang event is only a spontaneous eruption of energy from the vacuum state that occurs due to quantum uncertainty and the nature of that universal repulsion. A cosmic event horizon is like an inflating bubble in the void, which is the nature of the universe. The void is the empty background space that the universe is created within. The void is the ground state, or the vacuum state, from which all excited states arise (Penrose 2005, 28.4). In the sense of the holographic principle, those excited states are states of information defined on an event horizon. A state of information for the universe is defined on a cosmic event horizon.

It is instructive to examine thermodynamic principles more closely and how they relate to the holographic principle. Every possible event horizon is a possible state of the universe. That is where all the information for the universe is defined, with one bit of information per pixel on the screen. Those surfaces encode information, and are states of information. If that information is encoded with a binary code, like a sequence of 1's and 0's, then a surface with area A encodes a total number of bits of information $b=A/4\ell^2$, and the total number of ways to arrange all of that information is given by $N=2^b$. Each pixel encodes either a 1 or a 0, and there are N ways to arrange that information. In quantum theory, the entropy of any system in any distinct thermodynamic phase of organization is given by $S=k \log N$, where N is the number of distinct quantum states that give rise to the macroscopic appearance of the system in that distinct phase of organization. The entropy of a black hole measures all possible arrangements of information for any system. We conclude that the entropy of a black hole behaves like $S_{BH}=kA/4\ell^2$ (Penrose 2005, 27.10).

Thermodynamics is a description of how energy tends to flow from a hotter to a colder body, as a hotter body radiates away more heat. Thermodynamics is also a description of how information in any system becomes organized into distinct thermodynamic phases of organization, which gives rise to a distinct macroscopic appearance of that system. The distinct macroscopic appearance arises from the way information is organized at a microscopic level. For example, a system of water molecules can become organized into the macroscopically distinct phases of organization of either a gas or a liquid. A gas of hot water vapor can condense into liquid water if the temperature is lowered, which is called a phase transition. A system of water molecules that condenses into liquid water has less entropy than the system in the form of hot water vapor, since the number of quantum states that corresponds to the system in the form of liquid water is less than the number that corresponds to the form of water vapor. The entropy of a gas of hot water vapor is higher than the entropy of liquid water, since there are more distinct quantum states for the water molecules in that phase that give the same macroscopic appearance.

How is entropy related to the flow of energy? The entropy of any system is a measure of disordered information, and is inherently related to disordered kinetic energy. By a system, we mean some distribution of matter and energy that occupies a region of space. A gas of hot water vapor is such a system. If we put a bunch of water molecules inside a box, each water molecule has a certain position within the box, and moves with a certain velocity, which defines a microscopic state for the water molecules inside the box. We know from quantum theory that those position states and velocity states are quantized, and must take on discrete values. Those discrete values arise from probability amplitudes, and reflect the number of distinct wavelengths that can fit into the length, L, of the box. Momentum is quantized as $p=h/\lambda$, and the requirement of an integral number of wavelengths inside the box, $L=n\lambda$, quantizes momentum as $p=nh/L$, where n is an integer.

If the velocities of the molecules are great enough, that system of water molecules is in the macroscopic form of hot water vapor. As the water molecules collide into each other, they tend to scatter into random directions of motion. The second law says this gas of hot water vapor will come into thermodynamic equilibrium when its entropy is maximal, and the motion of the water molecules is as random as possible. Maximal entropy for this system occurs when the water molecules randomly move through all possible positions within the box, and randomly move with all possible velocities. The only constraint on the system is the volume of the box, and the temperature of the gas of water vapor. That temperature is a measure of the average kinetic energy of the water molecules. Each direction of motion for each water molecule is a degree of freedom that contributes to the amount of kinetic energy carried by that molecule, and defines temperature in terms of kinetic energy. If a water molecule moves with an average velocity v in some direction, it carries an average amount of kinetic energy KE=½mv², which by definition is set equal to KE=½kT, where T is the absolute temperature. If the total number of quantum states that gives rise to the macroscopic appearance of a hot gas of water vapor inside a box at temperature T is N, then the entropy of that gas is S=k*log*N (Penrose 2005, 27.3).

If the temperature is lowered, that gas of hot water vapor can condense into liquid water. How does this happen? The water molecules move around inside the box, collide with each other, and scatter off each other. But the water molecules also attract each other due to their electromagnetic energy of attraction, which arises from an uneven distribution of electric charges in space. Positive electric charge is located at an atomic nucleus, and negative electric charge on the electrons. Positive charges attract the negative charges. That is the force that holds a molecule together, but is also the force of attraction between different molecules. There is always a balance between the average amount of kinetic energy any molecule carries and its electromagnetic attraction to the other molecules. If there is too much kinetic energy, which is reflected by a high temperature, the molecules cannot bind together, since their average velocities are greater than an escape velocity. That escape velocity is determined when total energy E=KE+PE=0. If the temperature is lowered, the average velocity may fall below that escape velocity, and the molecules can bind together. The critical point occurs when the average velocity is equal to the escape velocity, and defines the critical temperature at which the phase transition occurs.

The concept of escape velocity is valid for both microscopic and macroscopic bound states. When an electron binds to a proton in a hydrogen atom, the electron's velocity must fall below escape velocity for the bound state to form. When two protons fuse together to form an atomic nucleus, the velocity of the protons must also fall below escape velocity. As any two particles bind together, the velocity of those particles must fall below escape velocity. If the velocity is initially greater than escape velocity, the only way that binding is possible is if some of that kinetic energy is radiated away. That energy is typically radiated away as a photon of electromagnetic radiation. As an electron binds to a proton, a photon is radiated away. As two protons fuse together, a photon is radiated away. The only reason the sun burns bright and shines with the energy of electromagnetic radiation is due to photons radiated away as protons fuse together inside the sun. Protons fell together from the big bang event under the influence of gravitational attraction to form the sun. As any bound state forms, energy is radiated away, which is the only way the velocity of the particles that fall together can fall below escape velocity.

It is instructive to examine how bound states form. The best example is a hydrogen atom, which is an electron bound to a proton under the influence of the electromagnetic force. The total energy of that system is given by E=KE+PE. If the total energy is positive, the system is unbound, and if negative, the system forms a bound state. All possible energy levels of the bound state are quantized. Quantization of energy follows from quantization of momentum, since p=h/λ. An integral number of wavelengths is required to fit into the circumference of any bound state orbit. If that orbit is circular with a radius r, the lowest energy level is defined by λ=2πr, which defines the ground state. The electron is not allowed to radiate away all of its kinetic energy and collapse at rest on top of the proton, since that would imply an infinite amount of energy by the uncertainty principle $\Delta p = h/\Delta r$. If there is no uncertainty in position, then there is infinite uncertainty in momentum. A point particle occupies no space. At this level, quantum uncertainty in position is the only reason a hydrogen atom occupies space (Feynman 1963, I 38-6). At a deeper level, point particles are impossible, since they imply infinite energy (Susskind 2008, 331).

The conventional way to understand the composite behavior of bound states is with the concepts of thermodynamics. An isolated point particle tends to move with uniform motion, but when two particles approach each other, they tend to scatter off each other due to the forces the particles exert on each other. In quantum field theory, the force exerted between two particles is understood as an exchange of another force particle between the two particles. For example, two electrons repel each other due to a photon exchange, and an electron and proton attract each other due to the exchange of a photon.

In this view of things, the nature of the electromagnetic force is conceptualized as due to the exchange of photons between charged particles. Kinetic energy is an aspect of the motion of particles, and potential energy is an aspect of the exchange of force particles.

But particles do not just scatter off each other as they exchange force particles with each other. Particles can also bind together into bound states. The simplest example to consider is a hydrogen atom, which is an electron bound to a proton. The total energy of the system is given by $E = KE + PE$. In non-relativistic physics, we express the electron's kinetic energy in terms of its mass, m, and its velocity, v, as $KE = \frac{1}{2}mv^2$, and its potential energy, as it moves relative to the proton at a distance of separation r, as $PE = -e^2/r$, where e is the charge of the proton. The negative sign indicates an attractive force. If the electron has too much kinetic energy, reflected by $E > 0$, the electron is unbound, and can only scatter off the proton. If $E < 0$, the electron is bound to the proton. The condition $E = 0$ defines escape velocity, or the velocity with which the electron must move at a distance r from the proton for the electron to just escape as it moves away from the proton. In this example, $(v_{esc})^2 = 2e^2/mr$. An electron with escape velocity has just enough kinetic energy to escape away from the proton, since $v_{esc} \to 0$ as $r \to \infty$.

In quantum theory, the easiest way to calculate the quantized energy level of a bound state orbital is to use the relationship $p = h/\lambda$ between momentum, p, and wavelength, λ. If the electron orbits the proton in a circular orbital of radius r, the requirement that an integral number of wavelengths fit into the circumference of the orbital, $n\lambda = 2\pi r$, gives the allowed quantized levels in terms of an integer n.

If we express kinetic energy in terms of momentum, $p = mv$, and use this relation, $p = nh/2\pi r$, then the total energy E is expressed only in terms of the radius r. This is equivalent to the statement that angular momentum is quantized as $pr = nh/2\pi$. This allows us to write $E = n^2h^2/8\pi^2mr^2 - e^2/r$. If we look for the minimum value of E as a function of r, as determined by $dE/dr = 0$, then for any value of n, we find the allowed quantized energy levels as $E_n = -2\pi^2me^4/n^2h^2$. This result gives all the Bohr orbitals (Feynman 1963, 38-6).

The ground state orbital is defined by n=1. The possibility n=0 is excluded, since it requires an infinite amount of energy.

There is nothing special about a hydrogen atom, except that it is an easy system to solve analytically for the energy levels. The same kind of analysis applies to any macroscopic system that forms as a bound state of point particles, except for the complexity of the mathematics involved, which may not allow for an analytic solution. The value of a thermodynamic discussion is to simply the math. With a thermodynamic analysis, we do not need to know the microscopic details of how every particle within a bound state moves relative to every other particle in the bound state. But thermodynamics tells us something very important about how any bound state forms. A bound state of point particles, whether microscopic or macroscopic in nature, can only form from a collection of point particles that are initially unbound to each other if kinetic energy is radiated away as the bound state forms. If an electron that is initially unbound to a proton approaches that proton, the bound state of a hydrogen atom can only form if kinetic energy is radiated away from that system in the form of a photon. The radiation of a photon away from the system of an electron and a proton carries away kinetic energy, which is the only way the velocity of the electron can fall below escape velocity. That is the only way the bound state can form. A photon must be radiated away.

The same kind of thing happens when liquid water freezes into ice. The water molecules attract each other due to an uneven distribution of electric charges in space. The force of attraction between water molecules is electromagnetic in nature, just as the force of attraction between an electron and a proton. The only difference is the electromagnetic force is screened out over long distances due to the overall electrical neutrality of the system of molecules. At shorter distances, there is a net electromagnetic attraction between molecules due to an uneven distribution of electric charges in space, which arises as the electric charges in the system are arranged in such a way as to minimize electromagnetic energy, just like the minimization of energy that occurs in the hydrogen atom. An uneven distribution of electric charges in space is a direct consequence of the exclusion principle, since only two electrons with opposite directions of spin can fill each atomic orbital, and also reflects that the negatively charged electrons orbit the positively charged atomic nuclei that compose the water molecules.

On average, two water molecules can only bind together if their velocities fall below an escape velocity determined by $E=KE+PE=0$. If the molecules have greater than escape velocity, they cannot bind together. If the molecules have less than escape velocity, they can bind together. There is a tendency for bound state pairs to form, then triplets to form, then quadruplets to form, and so on, until the entire system of liquid water binds into the bound state of frozen ice. But just like for the bound state of an electron and a proton, if there is too much kinetic energy in the system, the bound state will not form. The only way the bound state can form is if kinetic energy is radiated away in the form of photons. For a system of water molecules in the form of frozen ice, energy is minimized if the molecules taken on nearly fixed positions relative to each other, but this is only possible if kinetic energy is radiated away, and the average velocity of the molecules falls below an escape velocity determined by their attraction to each other.

As liquid water freezes into ice, kinetic energy is radiated away from the system into the environment in the form of heat. That radiation of heat is in the form of infrared photons. The radiation of heat occurs at a constant temperature, which is called the freezing point of water. Temperature is intrinsically related to kinetic energy. The absolute temperature of the system of water molecules is defined in terms of the average kinetic energy of each molecule for each direction of motion as $<KE>=\frac{1}{2}kT$. The critical temperature of the freezing point of water is defined by an escape velocity, at which the molecules have a tendency to bind together.

If there is too much kinetic energy, the system of water molecules takes the form of liquid water. As the temperature of the environment is lowered, heat is radiated away from the system into the environment, and temperature of the system falls. The phase transition from liquid water into ice occurs at the critical temperature, as the average velocity of the molecules approaches escape velocity, and they have more of a tendency to bind together. For the phase transition to go forward, more kinetic energy must be radiated away at the critical temperature, in order that the average velocity falls just below escape velocity. Heat must be radiated away even at the critical temperature for the phase transition to go forward. The amount of heat that must be radiated away at the critical temperature for liquid water to freeze into ice is called the latent heat. That heat is radiated away into the environment in the form of infrared photons.

The freezing of liquid water into ice is an example of coherent organization that develops as a phase transition occurs, and a collection of particles bind together into a bound state. Frozen ice is more organized than liquid water since the water molecules take on a more orderly arrangement of their positions in space. Coherent organization develops as the water molecules bind together. But the more highly organized arrangement of water molecules in frozen ice can only develop from liquid water if heat is radiated away from the system into the environment as the phase transition occurs. The radiation of heat carries away disordered kinetic energy, which allows the phase transition to go forward.

Thermodynamics expresses the concept of how highly ordered any arrangement of particles is within any bound state with the idea of entropy. The entropy of any system is a measure of hidden information, which is typically expressed in terms of bits of information. A bit of information is like a switch than can only assume the on or off position, and so encodes information in a binary code of 1's and 0's. A bit of information is like a question that can only be answered yes or no. The entropy of a system measures the number of such questions that can be asked about the microscopic details of the system, but are hidden in the macroscopic appearance of that system. In quantum theory, those microscopic details are specified in terms of quantum states. If the number of quantum states that give rise to the same macroscopic appearance of a system is called N, the entropy of the system, as it takes on that macroscopic appearance, is defined as $S=k\log N$, where k is Boltzmann's constant. Entropy is a measure of hidden information.

By a system, we typically mean some distribution of matter and energy in some region of space. A collection of point particles that bind together into a bound state is such a system. For a system of water molecules, the system takes on a lower entropy in the form of frozen ice than in the form of liquid water, since the molecules assume a more highly ordered arrangement of positions in space. In the sense of quantum theory, there are fewer microscopic quantum states that give rise to the macroscopic appearance of frozen ice than to the macroscopic appearance of liquid water for the same collection of water molecules in the same region of space. If we put those water molecules inside a box, and heat liquid water until it boils and evaporates into steam, there are more microscopic quantum states that give rise to the macroscopic appearance of steam inside a box than to the macroscopic appearance of liquid water, and so the entropy of the system in the form of water vapor is greater than the entropy of the system in the form of liquid water or frozen ice. In the same way that liquid water can only freeze into ice if disordered kinetic energy is radiated away as heat, liquid water can only evaporate into water vapor if heat is applied to the system. Heat must be applied at the boiling point of water for liquid water to evaporate into water vapor. The temperature of the system measures the amount of disordered kinetic energy of the water molecules. A change in phase of the system is a change in the way the system is coherently organized, which reflects the different possible ways in which the system of water molecules can form a bound state.

A coherently organized bound state can only form if heat is radiated away from the system. Heat is radiated away as infrared photons, which carry disordered kinetic energy away from the system. That radiation is the only way the velocity of the particles that are binding together can fall below an escape velocity. That is the only way a more highly ordered phase of the system can develop. The entropy of the system locally decreases as a consequence of that development of coherent organization. But that local decrease in entropy is only possible if there is a global increase in entropy. That global increase in entropy occurs due to the disordered kinetic energy radiated away into the environment. The phase transition only occurs if heat is radiated away into the environment.

The second law of thermodynamics states the total entropy of any system and its environment tends to increase over the course of time. A local decrease in the entropy of a system is possible if there is a global increase in the amount of entropy in the system and its environment. That global increase in entropy occurs due to disordered kinetic energy radiated away into the environment. The other way to state the second law is that heat tends to flow from a hotter to a colder object. The hotter object radiates away more heat. Heat is a form of disordered kinetic energy, typically in the form of electromagnetic radiation, like infrared photons. If the environment is colder than the system, heat flows from the system into the environment, since a hotter object radiates away more heat than a colder object. As the system radiates away heat into the environment, the system can become more ordered, but only if the environment becomes more disordered.

Entropy is a measure of the number of microscopic arrangements of the system that give rise to the same macroscopic appearance. We understand the nature of those microscopic arrangements as quantum states. At the most fundamental level possible, those quantum states encode information as bits of information, just like a switch that encodes information in a binary code. A switch that only assumes the on or off position encodes information in a sequence of 1's and 0's. At the most fundamental level possible, information is encoded in precisely this way. This is exactly what black hole entropy, and the coherent entropy bound, tells us about how information is encoded in the universe.

Macroscopic systems can undergo phase transitions, and form bound states, which are macroscopically distinct phases of organization. A gas of hot water vapor forms a bound state as it condenses into liquid water. This happens naturally as the temperature of the environment is lowered, and the system of hot water vapor radiates away heat into the environment. Heat tends to flow from the hotter to the colder object, and as the temperature of the environment is lowered, heat is radiated away into the environment. The heat radiated away is disordered kinetic energy, radiated away in the form of infrared photons. As heat is radiated away, the velocity of the molecules decreases, and they have a greater tendency to bind together under the influence of the attractive electromagnetic force between water molecules. As their velocities fall below an escape velocity, they tend to bind together into the bound state of liquid water. The bound state only forms if disordered kinetic energy is radiated away into the environment. The phase transition only occurs if heat is radiated away from the system. As the bound state forms, the total entropy of the water molecules and the environment tends to increase, due to the disordered kinetic energy radiated away into the environment (Penrose 2005, 27.7).

What happens if that system of water molecules gravitationally collapses into a black hole? A black hole is the system and its environment in its maximally disordered form, with maximal entropy. The black hole is a maximally disordered state that arises from maximal gravitational collapse. The black hole is the bound state that gives maximal entropy. Maximal entropy is a measure of the total number of possible arrangements of information for any system and its environment. The macroscopic appearance of a black hole is no more fundamental than the appearance a system takes on with the macroscopic appearance of a gas of hot water vapor or liquid water. What is fundamental?

If we understand the nature of entropy for a black hole, then we understand the fundamental microscopic level at which all information is quantized. Quantum gravity is about how information is defined at the fundamental level of quantized space-time. A black hole is the key that solves the puzzle. We already know from the holographic principle that information is encoded on the surface of the event horizon. The total number of bits of information encoded is $b=A/4\ell^2$, which gives the black hole an entropy of $S_{BH}=kA/4\ell^2$. That is the maximal entropy possible for any system holographically defined on the screen. Black hole entropy tells us about the maximal number of arrangements of information for any possible system. Black hole entropy is maximal since both the system and its environment are holographically defined on the screen. The holographic principle implies both system and environment are defined on the screen.

The covariant entropy bound (Bousso 2002, 19) establishes that the entropy of any region of space is bounded (in Planck units) by one-quarter of the surface area of the bounding surface of that space. This result follows from the idea of a light-sheet, which is constructed from light rays that emanate orthogonally from that surface and terminate at a focal point (Bousso 2002, Figure 4). The light-sheet is fundamentally related to the light cone (Penrose 2005, Figure 17.10).

The covariant entropy bound is fundamentally related to the idea of a Penrose diagram, which gives a graphical representation of the 'observable world' as observed from the point of view of any observer. That observable world is always limited by an event horizon for an observer in an accelerating frame of reference, due to the constancy of the speed of light (Penrose 2005, figure 27.16). In an exponentially expanding universe with a positive cosmological constant, referred to as de Sitter space, that observable world is limited by a cosmic event horizon (Bousso 2002, figure 10).

Hawking also calculated the temperature of the event horizon of a black hole as observed by a distant observer, and found that $kT=hc^3/16\pi^2 GM$. The other way to express this is as $kT=hc/8\pi^2 R$, where R is the radius of the event horizon. In thermodynamics, the value $E=kT$ is a measure of the amount of disordered kinetic energy in any system per degree of freedom, and defines temperature in terms of energy. A degree of freedom is like a bit of information defined on the screen. Hawking found the smallest bit of information that can escape away from a black hole is a photon with energy $E=hc/8\pi^2 R$. The photon is radiated away in the sense of thermal blackbody radiation. The encoding of information on the event horizon is inherently related to the separation of virtual pairs at the horizon.

The temperature of a solar mass black hole is about 10^{-7} degrees Kelvin. But the formula $kT=hc/8\pi^2 R$ also applies to a cosmic event horizon. The radial size of the universe at the time of the big bang event was about a Planck length, $R=\ell$, which gives the big bang event a temperature of about 10^{32} degrees Kelvin. The current temperature of the universe, measured with the observed spectrum of microwave radiation left over from the big bang, indicates a temperature of 2.7 degrees Kelvin. The universe has cooled since the big bang since it has expanded. If the universe continues to expand indefinitely, its temperature will approach absolute zero as its size approaches infinity, called the heat death of the universe. In an exponentially expanding universe with dark energy, there is always a cosmic event horizon, where the universe at that point appears to expand at the speed of light. Eventually all other matter and energy in the universe will cross the cosmic event horizon, and the universe will only contain dark energy, and nothing else. The current measured rate of exponential expansion of the universe indicates the size of the cosmic event horizon is about 15 billion light years, which gives it a temperature of about 10^{-30} degrees Kelvin. The second law ultimately describes how heat flows from the big bang event to the heat death of the universe. The big bang event is the hottest state of the universe, and a maximally inflated cosmic event horizon is the coldest possible state.

Inflationary cosmology tells us the total energy of the universe is zero, since the universe arises from the vacuum state as a spontaneous eruption of energy, due to the virtual creation of particle-antiparticle pairs. Those virtual pairs are created out of nothing and normally annihilate back into nothing, with a total energy that adds up to zero. Virtual pairs appear to separate at the cosmic event horizon, as the antiparticle appears to cross the horizon. That separation is how a universe of matter is created. How can the total energy of a universe of matter add up to zero? The answer is gravitational attraction. The negative potential energy of gravitational attraction cancels out all forms of positive energy, like mass energy and kinetic energy. Even the dark energy that is responsible for the exponential expansion of the universe is canceled out by gravity. Everything ultimately adds up to zero. The holographic principle explains how all the information for the universe is encoded on the surface of the event horizon. That encoding of information is inherently related to the separation of matter from antimatter at the horizon.

The holographic principle is the only known way to unify relativity theory with quantum theory, and unify the equivalence principle with the uncertainty principle. It is explicitly demonstrated in string theory, which is our best unified theory. String theory is the only consistent theory we have that quantizes gravity (Greene 1999, 135). String theory has a 'point particle' kind of description, which is the description of vibrating strands of string, and is similar to a quantum field description. But string theory also has a 'dual' description, which is the viewing screen description (Susskind 2008, 290). The viewing screen is an event horizon, or a two dimensional surface that encodes pixilated bits of information. Each fundamental pixel on the viewing screen encodes a quantized bit of information. The images of the things that we observe in our usual three dimensional world, like images of point particles, are holographically projected from the viewing screen to a central point of view. The probability of measuring those measurable images is inherent in the probability amplitudes that are calculated with a quantum field theory.

Unified theories like string theory assume the existence of an empty background space. It cannot be stressed strongly enough that this empty background space is the nature of the vacuum state, or the ground state from which all excited states arise. Those excited states of information are defined on the surface of an event horizon, as observed by an observer at the central point of view. This empty background space is not the same as our usual 3+1 dimensional space-time, which only has a holographic reality. String theory is only consistently defined as a quantum theory in a ten dimensional background space, due to an anomaly in the quantized spectrum of energy levels in any other dimensionality, which invalidates Lorentz invariance (Susskind 2008, 339). But these extra dimensions are exactly what is needed to unify all the fundamental forces with gravity.

String theory unifies all the fundamental forces through compactification of extra dimensions. The laws of the universe arise from the symmetry of empty space, like conservation of momentum that arises from translational symmetry of empty space, and conservation of angular momentum that arises from rotational symmetry of empty space. The symmetry of empty space expressed by relativity theory is the principle of equivalence (Greene 1999, 61), which expresses that all forces are inherently geometrical in nature, and are equivalent to accelerations. Every force is equivalent to an accelerating frame of reference in empty space. An event horizon always arises from the point of view of the observer at the center of that frame of reference, and is as far as the observer can see things in space, due to the constancy of the speed of light. A cosmic event horizon is a surface where the universe appears to expand at the speed of light, as observed from the central point of view of that sphere. Since nothing can travel faster than the speed of light, that surface is as far out as that observer can see things in space.

String theory unifies the fundamental forces of the universe with gravity through the compactification of extra dimensions. Electromagnetism is unified with gravity with compactification of an extra fifth dimension, and the strong and weak forces with compactification of another extra five dimensions (Susskind 2008, 339). The symmetry inherent in unification is expressed by the principle of equivalence. All the fundamental forces are equivalent to an accelerating frame of reference, and express the equivalence of all points of view in empty space. An implication of the principle of equivalence is the effects of all forces disappear for an observer in a state of free fall through empty space. Unification expresses this symmetry of empty space.

In relativity theory, a force is always equivalent to an acceleration. A path through space followed by a point particle is determined by a geometrical principle, which is the path of least action. In relativity theory, action is equivalent to proper time (Penrose 2005, 17.8, 20.1). We imagine particles carry clocks with them, and the amount of time that passes in the particle's rest frame is its proper time. Particle motion is determined by the path that minimizes the action, which is also the path that maximizes the amount of proper time. Time appears to run more slowly on a clock in motion due to time dilation, but appears to run more rapidly for an observer in an accelerating frame of reference (Greene 1999, 37, 74). For example, time appears to run faster on a clock above the surface of the earth as observed by an observer at the earth's surface, due to the force of gravity.

It is instructive to briefly review these two effects. The first effect is time dilation. Imagine that the particle's clock consist of two mirrors, and a photon bounces back and forth between the mirrors (Greene 1999, 39). Each bounce of the photon off a mirror is a tick of the clock. If the mirrors are separated by a distance L in the y-direction, the amount of time that passes by with each tick as observed in the particle's rest frame, and as the photon moves at the speed of light, is $\Delta\tau=L/c$. From the point of view of another observer, that time interval is Δt. If the particle appears to move in the x-direction with velocity v, then in a time interval Δt, the particle will appear to move a distance $\Delta x = v\Delta t$. The total distance, D, the photon appears to move as it makes one tick is given by the Pythagorean theorem $D^2 = L^2 + (\Delta x)^2 = c^2(\Delta\tau)^2 + v^2(\Delta t)^2$. The speed of light is a constant, so from the point of view of the other observer $D = c\Delta t$. It is a simple matter to rearrange these terms, which results in the effect of time dilation $(\Delta t)^2 = (\Delta\tau)^2/(1 - v^2/c^2)$. From the point of view of the other observer, time appears to run more slowly.

The second effect arises with accelerated motion. Imagine that a spaceship accelerates through empty space with an acceleration rate a=g. A clock is placed in the front of the spaceship and an identical clock in the back of the spaceship, separated by a distance x. Imagine the clock in the front of the spaceship emits a flash of light every $\Delta\tau$ seconds. An observer at the back of the spaceship measures the arrival of each flash of light. That observer compares the time interval Δt between the arrival of each flash of light with the time interval recorded on the clock in the back of the spaceship, which is $\Delta \tau$, since the clocks are identical. The flash of light appears to arrive early, since the spaceship accelerates while the flash of light travels at the speed of light from the front clock to the back clock, and so the observer at the back of the spaceship observes that the front clock appears to run fast. The easiest way to calculate how much the front clock appears to run fast is to use the Doppler effect (Feynman 1963, II 42-9), which gives $\Delta t=\Delta \tau/(1+gx/c^2)$.

Another way to see this is to use the equivalence principle. Consider a photon that falls in a gravitational field. The gravitational potential energy of an object of mass m at a height x above the surface of the earth is given by PE=mgx. If a photon with energy E=hv falls from a height x down to the surface of the earth, that photon gains an amount of potential energy ΔE=mgx. That photon has a 'gravitational mass' specified by $E=mc^2=hv$, and so $\Delta E=hvgx/c^2$. As the photon falls in the gravitational field its frequency increases by an amount $\Delta E=h\Delta v$, which gives $\Delta v=vgx/c^2$. The energy of the photon at the surface of the earth relative to its energy at a height x above the earth's surface is given by $E'=hv'=h(v+\Delta v)=hv(1+gx/c^2)$, which is the same as $\Delta t=\Delta \tau/(1+gx/c^2)$ if we identify the frequency of vibration with an inherent rate of oscillation as $v=1/\Delta \tau$.

The principle of equivalence tells us there is no way to distinguish the effects of a gravitational field from an accelerating frame of reference. The acceleration due to gravity on the surface of the earth is $g=GM/R^2$, where M is the mass and R is the radius of the earth. If we have a clock on the surface of the earth and an identical clock at a height x above the surface of the earth, the clock at the higher elevation appears to run faster by an amount $\Delta t=\Delta \tau/(1+gx/c^2)$. The clock at the higher elevation appears to run faster due to its equivalence to a clock in an accelerating frame of reference.

How do we discover the action principle from these two effects? Einstein tells us to look at the proper time interval, which is the amount of ordinary time that passes in the particle's rest frame. Einstein tells us the particle follows a geometrical path through space-time that maximizes the proper time relative to all nearby paths. That path is like the shortest distance between two points in a curved space-time geometry (Zee 2003, 79). If that particle moves in the earth's gravitational field at a height x above the surface of the earth, the time interval measured on the particle's clock as observed by an observer at the earth's surface is $\Delta t=\Delta \tau/(1+gx/c^2)$. But if that particle moves with velocity v, we also have to include the effects of time dilation, which gives $\Delta \tau=\Delta t(1+gx/c^2)(1-v^2/c^2)^{1/2}$. In the non-relativistic limit we can approximate $\Delta \tau=\Delta t(1-\frac{1}{2}v^2/c^2+gx/c^2+...)$. If we define the particle's action in terms of its mass m and its proper time as $\Delta S=-mc^2\Delta \tau$, then $\Delta S=\Delta t(\frac{1}{2}mv^2-mc^2-mgx)=(KE-PE)\Delta t$. The potential energy includes the mass energy and the gravitational potential energy as $PE=mc^2+mgx$. Einstein tells us the particle will follow a path through space-time that maximizes the proper time interval, or minimizes the action. Kinetic energy arises from time dilation, and potential energy arises in an accelerating frame of reference. Both kinds of energy are purely geometrical in nature.

Even the $1/R^2$ force law, as in $g=GM/R^2$, is geometrical in nature, and arises from the amount of particle flux that crosses the surface area of a sphere (Zee 2003, 27).

The path of least action minimizes the amount of kinetic energy expended by the particle on its path while it maximizes the amount of potential energy preserved on that path, as observed from the point of view of the observer of that motion. Kinetic energy arises from the effects of time dilation, while potential energy arises in an accelerating frame of reference. Time dilation and kinetic energy arise with all motion due to the constancy of the speed of light. Every force is characterized by potential energy and an accelerating frame of reference. The path of least action is the classical path (Feynman 1963, II 19-1), which allows us to recover the classical laws of motion, F=mg=ma, or a=g, which was our initial assumption. We have come full circle. We are back where we started!

How is this related to quantum theory? The classical path is the path of least action, but is only the most likely path in the sense of quantum probability. The particle can follow all possible paths, or trajectories x=x(t), but some are more likely than others. Quantum theory instructs us to calculate probability amplitudes $z(\theta)=\exp(i\theta)=\cos(\theta)+i\sin(\theta)$, where $i=\sqrt{(-1)}$ is the unit imaginary number. This remarkable formula is called Euler's formula. The phase angle θ is related to action as $\Delta\theta=2\pi\Delta S/h$, where h is Planck's constant. We are instructed to sum over all possible paths, and assign each path a probability factor $z(\theta)$ that depends on the action for that path. The probability amplitude $z(\theta)$ acts like a vector in the complex plane. That vector points in some direction relative to the real axis, with an angle θ. As we sum over all possible paths, those vectors tend to cancel each other out, since they tend to point in random directions. The only paths that do not tend to cancel out are those near the path of least action, which is the path that minimizes the amount of action (Penrose 2005, figure 26.4). The angle θ acts just like a phase angle in an interference pattern. When waves are in phase they add together, and when out of phase they cancel out. The path of least action is the stationary path that has the greatest phase reinforcement, and gives the largest quantum probability (Penrose 2005, 26.6).

How is this related to a quantum field? A quantum field amplitude is expressed as a sum over plane waves, $\Psi(x,t)=A\exp(2\pi i[x/\lambda-\nu t])$, where A is a wave amplitude, λ is a wavelength, and ν is a frequency. This describes a wave oscillation thanks to Euler's formula. The sum over all possible wavelengths and frequencies describes any possible wave. These wave oscillations are describable in terms of a phase angle, just like the probability factor $z(\theta)$, where $\theta=2\pi(x/\lambda-\nu t)$. This result also follows directly from the non-relativistic action for a point particle $\Delta S=(KE-PE)\Delta t=(\frac{1}{2}mv^2-PE)\Delta t=(p\Delta x-E\Delta t)$, where p=mv, v=$\Delta x/\Delta t$, and E=KE+PE. We can write $\Psi(x,t)=A\exp(2\pi i[px-Et]/h)$, and make the quantum correspondence that $E=h\nu$ and $p=h/\lambda$. This is still only geometry.

Quantum theory is expressed as a sum over all possible paths. Each path is weighted with a probability factor that depends on the action $\Psi(x,t)=\exp(2\pi iS/h)$. For point particle motion the action is written as $\Delta S=p\Delta x-E\Delta t$. The only thing that makes quantum field theory more complicated than ordinary quantum mechanics is the closed loops of virtual particle-antiparticle pairs that arise within the path of any particle (Zee 2003, 56).

The probability factors embody the uncertainty principle. As the particle moves on a path from an initial position to a final position over the course of an interval of time, the probability factors describe wave motion. If we want to localize the particle within some distance interval Δx, we have to scatter light off the particle. The wavelength of the light determines how much momentum is carried by the light wave as $p=h/\lambda$. As the light wave scatters off the particle, there is an exchange of momentum. If we want to localize the particle within a distance interval $\Delta x=\lambda$, then an amount of momentum $\Delta p=h/\Delta x$ must be exchanged, which is the uncertainty principle (Feynman 1963, I 37-11).

The most likely path in the sense of quantum probability is the path of least action, which is the path that minimizes the expenditure of kinetic energy while maximizing the preservation of potential energy, as observed by the observer of that motion. The path of least action is also the path that maximizes proper time. As observed from the point of view of the observer, kinetic energy arises from the effects of time dilation, while potential energy arises in an accelerating frame of reference. The classical path is only a geometrical statement about the most likely path the particle can follow. In the sense of relativity theory, that path maximizes the amount of proper time, and is like the shortest distance between two points in a curved space-time geometry. That space-time geometry is curved purely due to the effects of time dilation, which reflects the principle of relativity, and the effects of an accelerating frame of reference, which reflects the principle of equivalence. The closed loops of virtual particle-antiparticle pairs that arise within any path reflect the uncertainty principle. The whole thing reflects geometry.

There is an alternative way to understand the sum over all possible paths, based on the idea of an eigenstate. A position eigenstate is interpreted as an allowed position of the particle as it moves on its path (Penrose 2005, 21.10). In this view of quantum physics, every event is a decision point where the quantum state of potentiality branches into alternative paths. The sum over all possible paths can be rewritten as a sum over all possible position eigenstates. The total probability amplitude that specifies the probability with which a point particle can propagate between two points can be written as a sum over all possible paths, or as a sum over all possible position eigenstates. In this view of things, an event is a decision point where the path branches into all possible alternative paths. The path of least action is most likely in the sense of quantum probability.

Some may argue that this procedure is not well defined mathematically, since there apparently are an infinite number of position eigenstates. We now know this objection is false. Our observations show us that we live in a universe with a positive cosmological constant. Observation shows the universe is exponentially expanding. In an exponentially expanding universe, every observer is surrounded by a cosmic event horizon, which is a spherical surface where the universe at that point appears to expand at the speed of light. Due to the constancy of the speed of light, an observer can only see things as far out as the cosmic horizon. Everything beyond a cosmic horizon is hidden from the observer. The argument for black hole entropy also applies to a cosmic horizon. The covariant entropy bound (Bousso 2002, 43) indicates the amount of information that can be encoded in such an exponentially expanding universe is finite. For a cosmic horizon at about 15 billion light years, about 10^{123} bits of information can be encoded. This proves the number of position eigenstates in such a universe is also finite (Greene 2011, 38).

To construct a quantum state of potentiality requires that we take a sum over all possible paths (Penrose 2005, 26.6). The sum over all paths for point particle motion results in a quantum field theory. Any quantum field theory amplitude, $\Psi(x,t)$, is a probability amplitude that specifies the probability that the point particle can be measured at position x at time t. Quantum field amplitudes are calculated with a sum over all possible particle paths. The path of least action is the most likely path in the sense of quantum probability.

We measure a particle-like behavior of the point particle when we measure its position at some moment of time (Susskind 2008, 80). Quantum field amplitudes also exhibit wave-like behaviors due to the sum over all possible paths. Those wave-like behaviors include phenomena like interference patterns. We measure a wave-like behavior when we measure the interference pattern (Susskind 2008, 78).

The difficulty in constructing a unified theory is how to unify the principle of equivalence with the uncertainty principle. All unified theories assume the existence of an empty background space, which is the nature of the vacuum state, or the ground state from which all excited states arise. That empty background space is not the same as the usual 3+1 dimensional space-time we are familiar with and observe, which only has a holographic kind of reality. All unified theories are inherently holographic in nature. All the fundamental bits of information for the form of anything observed in the world are holographically encoded on an event horizon, with one quantized bit of information per pixel on the screen. The viewing screen description is the more fundamental description. The viewing screen defines an excited state of information that arises from the vacuum state. The observation of the form of anything in the world is like the holographic projection of an image from the screen to a focal point of perception. Those images are animated over a sequence of events, just like the frames of a movie (Susskind 2008, 305).

String theory, like all unified theories, is inherently holographic in nature. String theory has a point particle like description that is similar to a quantum field theory, which is the description of vibrating strands of string. String theory also has something similar to the virtual particle-antiparticle pairs of quantum field theory, which are virtual string-antistring pairs that arise from the vacuum state due to quantum uncertainty. But there are no real point particles in string theory (Susskind 2008, 335). The fundamental description of string theory is the dual description of the viewing screen. In string theory, the viewing screen is an event horizon that encodes pixilated bits of information.

The viewing screen description is the more fundamental description, since that is where all the fundamental bits of information for the world are defined. A point particle, located at a point in three dimensional space at some moment of time, is like a holographic projection of an observable image from a two dimensional viewing screen to a point of view. The probability of observing that image at that point of view is determined by the probability amplitudes in string theory that describe the vibrating strands of string, which are similar to probability amplitudes calculated in a quantum field theory.

String theory demonstrates that there is no such thing as point particles that exist in some pre-existing space and time. The primordial nature of existence is the void. The world is holographically constructed within that empty background space. A viewing screen is an event horizon that encodes pixilated bits of information, and always arises from the central point of view of an observer. The images of things in the world are holographically projected to that point of view, and are animated over a sequence of events in the flow of energy, just like the frames of a movie.

Quantum theory tells us that the quantum state of the universe is a state of potentiality, which describes all possible paths that the universe can take in its dynamical evolution. Every event is a decision point where the quantum state branches into alternative paths. The path of least action is only the most likely path in the sense of quantum probability. The many world interpretation tells us that each path is actually taken, in the sense that an observer is always present for each path. In the sense of inflationary cosmology, each path of the universe is like a bubble in the void that inflates in size. That bubble is only an event horizon, which is a spherical surface that inflates in size, and always has an observer present at the central point of view of that bubble. A cosmic event horizon is a spherical surface where the universe at that point appears to expand at the speed of light. Since nothing can ever travel faster than the speed of light, the observer at the central point of view can only see things in space as far out as that horizon. Every observer has its own bubble, and is at the center of its own world. The quantum state of potentiality for the universe is a sum over all bubbles in the void, which is a sum over all surfaces.

The holographic principle explains how the surface of any bubble encodes information, acts like a holographic viewing screen, and projects images to the central point of view. Images are animated over events in the flow of energy, like the animated frames of a movie. The confusing aspect of consensual reality is each bubble shares information with other bubbles, which is the nature of the perceivable world that we share together. The mechanism by which information is shared is called quantum entanglement (Penrose 2005, 23.3). Any bubble has an observer present at the central point of view, but those surfaces are pixilated, and encode information. The quantum state of any bubble includes all possible ways that information can become encoded on all the different pixels.

A state of information for a bubble is defined by the way information is encoded. Every event is a decision point, which describes all the different ways in which information can become encoded on those pixels. An event is a decision point where the path branches. The path only branches due to all the different ways in which information can become encoded on those pixels. The other bubbles are described by their own states of information. Quantum entanglement describes how the different bubbles interact with each other, as bits of information tend to align together. That alignment allows the different bubbles to share information. What appears to happen in any bubble is connected to what appears to happen in other bubbles to the degree the bits of information in those different states of information interact with each other, align together, and share information.

The holographic principle explains the fundamental level at which all information is defined, but it also explains the source of all information, in the same way that inflationary cosmology explains the source of everything in the universe. The source of everything is the void. All excited states of information arise from the vacuum state. The void is the empty background space the universe is created within. The universe is like a bubble in the void. These theories tell us everything arises from the nothingness of empty space as a quantum fluctuation in the zero energy level of the void. We call that spontaneous eruption of energy from the void the big bang event. Information is only encoded on event horizons due to quantum uncertainty with that quantum fluctuation.

All information is encoded on surfaces of quantized space-time, which are event horizons in the sense of relativity theory, and define states of information. Information is pixilated on the surface. Each fundamental pixel on the screen encodes a quantized bit of information. A viewing screen is an excited state of information that arises from the void.

The encoding of pixilated bits of information on the event horizon only occurs as virtual particle-antiparticle pairs appear to separate at the event horizon. Virtual particle-antiparticle pairs are created out of nothing, and normally annihilate back into nothing within a short period of time, as specified by the uncertainty principle. But something very strange appears to happen at the event horizon, as observed by the observer at that central point of view. The virtual antiparticle appears to cross the event horizon, and is not observable to the observer at the central point of view, while the virtual particle appears to move toward the observer, and appears to become a real particle that is observable. Separation of matter from antimatter at the event horizon is how a universe of matter is created. Separation of virtual particles from virtual antiparticles at the event horizon creates a kind of holographic virtual reality, as virtual particles appear to become real (Susskind 2008, 171), and as information is encoded on the viewing screen.

Those bits of information tend to coherently align with each other, which makes the surface holographic. Coherent organization arises from alignment of information. Each distinct form of information is coherently organized, and tends to hold together over a sequence of events. Coherent organization is the nature of all the distinct things in the world that appear to hold together and self-replicate form. Coherent organization is the only way a distinct form holds together as a bound state of information, which allows for self-replication of form, while its behaviors are enacted over a sequence of events. In the sense of thermodynamics, any macroscopic form is organized within a coherent phase.

The nature of coherent organization, as bits of information tend to align together, arises from symmetry breaking. The symmetry that is broken is the symmetry of empty space. Symmetry breaking is how bound states of information form, which allows for self-replication of form. Energy flows in the sense of thermodynamics. In the usual quantum field theory description of point particles, the particles tend to randomly move around, and tend to scatter off each other in collisions, due to their kinetic energy. But particles also tend to bind together into bound states due to their potential energy of attraction.

In the viewing screen description, bits of information tend to randomly flip back and forth, which is the viewing screen analogue of kinetic energy. A pixel that encodes a bit of information in a binary code of 1's and 0's is like a switch that flips back and forth between the 'on' and 'off' position. Bits of information tend to align with each other due to quantum entanglement, which is the viewing screen analogue of potential energy. Spin networks (Penrose 2005, 32.6) demonstrate how this is possible. Quantized bits of information align together as entangled states of spin angular momentum add together. In the same sense as any other quantum theory, the amount of action that separates two events is given by $\Delta S=(KE-PE)\Delta t$, where KE arises as bits of information flip back and forth, and PE arises as bits of information tend to align with each other.

Alignment of information allows bound states of information to form, which is the nature of coherent organization that allows for the formation of animated forms of information that replicate their forms over a sequence of events. That alignment of information spontaneously emerges in the flow of energy, and breaks the symmetry of empty space.

Alignment of information arises from symmetry breaking. Rotational symmetry of empty space leads to conservation of angular momentum and quantization of spin angular momentum. If space-time geometry was 3+1 dimensional, only spin angular momentum of point particles would arise, like the spin ½ electron and spin 1 photon. String theory, like any unified theory, assumes the existence of an empty background space. The electromagnetic, strong and weak forces arise from the compactification of extra dimensions, which lead to 'gauge' symmetries (Greene 1999, 124, 374). Any compactified dimension is like another rotational symmetry. Multiple compactified dimensions lead to the encoding of quantized bits of information, but in larger rotational groups than ordinary spin (Greene 1999, 186, 205).

Unified theories also assume super-symmetry, which is a strange kind of symmetry. Any point in empty space is located with ordinary commuting numbers and anti-commuting numbers. Greene describes super-symmetry as "just as spin is like rotational motion with a quantum-mechanical twist, super-symmetry can be associated with a change in observational vantage point in a quantum-mechanical extension of space and time" (Greene 1999, 172). The encoding of information naturally arises with compactification of extra dimensions in an empty background space due to symmetry. Alignment of information arises as states of information become entangled, like entangled spin states. Entangled states are mathematically represented by the multiplication of states (Penrose 2005, 23.4). The rules of group theory describe how entangled states multiply together, and how spin states combine together (Zee 2003, 468). As entangled states add together, quantized bits of information tend to align together like little magnets, which allows for the formation of bound states of information on an event horizon, as observed from the central point of view. Each pixel on the screen encodes a quantized bit of information.

It is instructive to examine how information is encoded with the compactification of extra dimensions. The proto-typical example is the Kaluza-Klein mechanism (Zee 2003, 428). Relativity theory is defined in five space-time dimensions with a compactified fifth dimension. A compactified fifth dimension is rolled-up into a small circle. This is usually compared to the surface of a garden hose, which appears like a one dimensional line when examined from a long distance away, but appears two dimensional when examined up close (Greene 1999, 186). The compactified fifth dimension is rolled-up into a small circle at every point of our usual 3+1 dimensional space-time geometry. Einstein's equations are written in terms of the metric, which measures the curvature of space-time geometry, and is the nature of the gravitational field. But the metric also measures the amount of proper time that passes on the path of any particle through that space-time geometry. The path of least action maximizes the amount of proper time, and is like the shortest distance between two points in a curved space-time geometry (Zee 2003, 79).

The remarkable aspect of this procedure is that Maxwell's equations of electromagnetism naturally arise from Einstein's equations with the compactification of the fifth dimension (Zee 2003, 433). The electromagnetic field naturally arises from the components of the metric that describe how that space-time geometry curves into the fifth dimension. Even more remarkable is the nature of electric charge at any point in space-time arises from the compactification of the fifth dimension. Electric charge is nothing more than momentum directed in the compactified fifth dimension at every point of our usual 3+1 dimensional space-time geometry. Fifth dimensional momentum is quantized as $p=h/\lambda$, and can be directed in the positive or negative direction. If that compactified dimension has a radius of r, the requirement that an integral number of wavelengths fit into the circumference of that circle, $n\lambda=2\pi r$, quantizes momentum, and quantizes electric charge, as $p=nh/2\pi r$. The momentum quantized with the compactification of an extra dimension is similar to a spin variable, since $pr=nh/2\pi$ is just like spin angular momentum quantized in integral units at every space-time point. The concept of a spin network thus allows us to understand how information is encoded on the surface of an event horizon. Each pixel on the screen encodes a quantized bit of information. Alignment of information naturally arises as spin states become entangled and combine together. The whole thing is pure geometry.

The phenomena of quantum entanglement raises certain metaphysical questions about the nature of measurement and observation. These metaphysical questions are at the very center of how we understand quantum theory. There are those that wish to take the meta out of physics, but that is not possible. It is impossible to take the meta out of physics since it is impossible to take the observer out of physics. The very nature of physics would not exist without observation. The principle of equivalence is inherently based upon the nature of the observer. All the debate about the correct interpretation of quantum theory is about the nature of observation. The only way to take the meta out of physics is to take the observer out of physics, which is impossible. The only other option is to explain the nature of the observer with a physical theory, but that is equally impossible. Such a physical theory of the observer would give a physical explanation of the nature of consciousness, but no such physical explanation is possible. That is what the incompleteness theorems prove. The observer of a physical world can never be reduced to the way information is physically encoded or coherently organized in that physical world. The observer is always 'outside' of that physical world. The physical body of an experimenter is not the same as the consciousness that is present for that physical body.

After quantum theory was formulated, the standard interpretation of quantum theory was proposed, but nobody was very happy with this idea, and it continues to be hotly debated to this day (Penrose 2005, 29.1). The standard interpretation proposes that a measurement is a quantum state reduction (Penrose 2005, 22.1). The conceptual difficulty with the standard interpretation is that a quantum state reduction of entangled spin states violates Bell's theorem (Penrose 2005, 23.3; Shimony 2009, 11). Bell's theorem expresses the expectation that measurable variables physically separated far away from each other should behave independently of each other. Bell's theorem assumes that there is no possibility that the result of a measurement of one variable can effect the measurement of another variable that is physically separated by a large distance. Bell's theorem is explicitly violated in the measurement of physically separated entangled spin variables.

The violation of Bell's theorem in the measurement of physically separated entangled spin variables is a conceptual problem, since it is strong evidence that particles do not behave like independent entities after they become physically separated. This demonstrates the impossibility of a locally realistic interpretation of quantum theory. In other words "*no physical theory which is realistic and also local in a specified sense can agree with all of the statistical implications of Quantum Mechanics*" (Shimony 2009, 2).

This is how Shimony describes these conceptual difficulties: "Quantum nonlocality and Relativistic locality–may have less to do with signaling than with the ontology of the quantum state. Heisenberg's view of the mode of reality of the quantum state was–that it is *potentiality* as contrasted with *actuality*". He goes on to state: "the domain governed by Relativistic locality is the domain of actuality, while potentialities have *careers* in space-time (if that word is appropriate) which modify and even violate the restrictions that space-time structure imposes upon actual events. The peculiar kind of causality exhibited when measurements at stations with space-like separation are correlated is a symptom of the slipperiness of the space-time behavior of potentialities. This is the point of view tentatively espoused by the present writer, but admittedly without full understanding. What is crucially missing is a rational account of the relation between potentialities and actualities–just how the wave function probabilistically controls the occurrence of outcomes. In other words, a real understanding of the position tentatively espoused depends upon a solution to another great problem in the foundations of quantum mechanics–the problem of reduction of the wave packet" (Simony 2009, 31).

Shimony continues: "*something* is communicated superluminally when measurements are made upon systems characterized by an entangled state, but that something is *information*, and there is no Relativistic locality principle which constrains its velocity". He states: "A radical idea concerning the structure and constitution of the physical world, which would throw new light upon quantum nonlocality, is the conjecture–about the nature of space-time in the very small, specifically at distances below the Planck length (about 10^{-33} cm). Quantum uncertainties in this domain have the consequence of making ill-defined the metric structure of General Relativity Theory. As a result–basic geometric concepts like point and neighborhood are ill-defined, and non-locality is pervasive rather than exceptional as in atomic, nuclear, and elementary particle physics. Our ordinary physics, at the level of elementary particles and above, is (in principle, though the details are obscure) recoverable as the correspondence limit of the physics below the Planck length. What is most relevant to Bell's Theorem is that the non-locality which it makes explicit in Quantum Mechanics is a small indication of pervasive ultramicroscopic nonlocality. If this conjecture is taken seriously, then the baffling tension between Quantum nonlocality and Relativistic locality is a clue to physics in the small".

The natural way to understand the nature of quantum nonlocality is with the holographic principle of quantum gravity. Information is not encoded in 3+1 dimensional space-time, but on the two dimensional surface of an event horizon, as observed by an observer at the central point of view. For this formulation to make sense, there is one missing ingredient, which is the many world interpretation of quantum theory (Penrose 2005, 29.1).

The many world interpretation of quantum theory, as put forward by Hugh Everett, is considered too far-fetched and too radical an idea by many physicists. Everett was a student of John Wheeler at Princeton. Wheeler's other students included Richard Feynman, who discovered the sum over all paths formulation of quantum theory, and Jacob Bekenstein, who first calculated the entropy of a black hole, and which led to the discovery of the holographic principle. Even today, there's a split in the physics world about the correct interpretation of quantum theory (Penrose 2005, 29.2).

The many world interpretation is seen as the natural interpretation by those that accept it. Those that hold onto the standard interpretation see the flaws of that interpretation, and don't like it, but consider the many world interpretation as too far-fetched and too radical an idea. But there is no natural way to understand the holographic principle without it.

Quantum entanglement and the violation of Bell's theorem tell us that there really is no such thing as independent entities called point particles that exist in some pre-existing space and time. But we already knew that from the holographic principle. All the bits of information for a particle are encoded on the surface of an event horizon, as observed by the observer at the central point of view. It helps to deconstruct the holographic principle and identify exactly where our usual ideas about the nature of physical reality go wrong.

The simplest case to consider are two entangled spin ½ variables (Penrose 2005, 23.3). In quantum theory a spin 'up' eigenstate is designated by $\Psi=|\uparrow>$, and a spin 'down' eigenstate by $\Psi=|\downarrow>$. An arbitrary quantum state of spin is written as $\Psi=a|\uparrow>+b|\downarrow>$. The parameters 'a' and 'b' are probability amplitudes that specify the likelihood with which the spin ½ variable can be measured in the 'up' or 'down' spin eigenstates. That measurement is a quantum state reduction that reduces the quantum state to either $|\uparrow>$ or $|\downarrow>$.

Spin ½ can only be measured to be 'up' or 'down'. That is what quantum state reduction means. An arbitrary quantum state is like a probability distribution that says with a likelihood determined by the parameter 'a' the spin can be measured to be 'up', and with a likelihood determined by 'b' that the spin can be measured to be 'down'. Those are the only two possibilities. A spin ½ variable can only be measured to be 'up' or 'down', which means the probability amplitudes satisfy $a^2+b^2=1$. A spin measurement is a quantum state reduction that reduces the quantum state to either $|\uparrow>$ or $|\downarrow>$. That measurement requires a choice. Quantum theory says that nature makes her choices randomly. Even if the choice is made randomly, the probability distribution that is measured is not random, which allows for correlation of behavior between different measurements.

The classic 'thought' experiment is to take an unstable spin zero particle, like a pi-meson, and let it decay into two spin ½ particles. The two spin ½ particles move in opposite directions away from the initial position of the spin zero particle. Let's call those directions R and L. Diagrammatically:

L-side   (Spin ½)←←(Spin 0)→→(Spin ½)   R-side

The most general quantum state of a single spin ½ variable is $\Psi=a|\uparrow>+b|\downarrow>$. When two spin ½ variables interact with each other, their quantum states become entangled, which is mathematically expressed by the multiplication of those quantum states. But the two spins are constrained by the total amount of spin that arises from the pi-meson decay, which is zero. The entangled quantum state of that decay process, which describes a total spin of zero, is written as (Penrose 2005, 23.4):

$$\Psi=a|\uparrow R>|\downarrow L>+b|\downarrow R>|\uparrow L>$$

The total spin has to add up to zero since spin is a conserved quantum number, which arises from rotational symmetry and conservation of angular momentum. For the total spin to add up to zero, if the R-particle is 'up' then the L-particle must be 'down', and if the R-particle is 'down' then the L-particle must be 'up'. Those are the only two possibilities that add up to a total spin of zero. The two spin ½ particles move away from each other. Let's imagine that one travels to Mars and the other travels to Venus. A clever experimenter on Mars measures the direction of that spin as the particle passes by. There is another clever experimenter on Venus that measures the direction of that spin as it passes by. In the standard interpretation, a measurement is a quantum state reduction.

The quantum state can only be reduced to $|\uparrow R>|\downarrow L>$ or to $|\downarrow R>|\uparrow L>$. Those are the only two possibilities. A measurement is a quantum state reduction that chooses among these two possibilities. If the experimenter on Mars measures the particle spin that passes by to be 'up', then the experimenter on Venus must measure the particle spin that passes by to be 'down'. If the experimenter on Venus measures the particle spin that passes by to be 'up', the experimenter on Mars must measure the particle spin that passes by to be 'down'. The results are always correlated with each other since the quantum states are entangled.

Quantum entanglement indicates that different measurements, performed by different experimenters far away from each other, are correlated with each other. The results of independent measurements, separated far away from each other, are correlated due to quantum entanglement. Entangled quantum states are reduced together, no matter how far apart the things are that are measured. Those things can be measured on opposite sides of the universe, but as soon as one measurement is performed, the other is also determined.

Until the measurement is performed the quantum state is a state of potentiality. Once the measurement is performed the quantum state is reduced. What one experimenter actually measures is determined by what the other experimenter actually measures. It does not matter if the two experimenters are on opposite sides of the universe. The moment the experiment is performed on one side of the universe by one experimenter, the result of the experiment is also determined on the other side of the universe.

Einstein referred to this phenomena as 'spooky action at a distance', indicating the presence of a 'ghost'. He was absolutely right. That ghost is a presence of consciousness that perceives whatever appears to happen in its world. That ghost is always present at a point of view, while the images of its world play like movie images on a viewing screen.

The holographic principle explains the subjective nature of reality. There is no such thing as objective reality. Susskind describes this state of affairs as: "The objective reality of points of space and instants of time is on its way out, going the way of simultaneity, determinism, and the dodo. Quantum gravity describes a much more subjective reality than we ever imagined" (Susskind 2008, 8). If reality was objective in nature, information in 3+1 dimensional space-time could be encoded on a three dimensional lattice of quantized space, referred to as *voxels* (Susskind 2008, 295). But information is not encoded in three dimensional space. Information is pixilated, and is encoded on the two dimensional surface of an event horizon, as observed by the observer present at the central point of view of that surface. The encoding of information arises purely from the principle of equivalence, which expresses the equivalence of all points of view in empty space, and the uncertainty principle, which explains how something is created from nothing as virtual particle-antiparticle pairs appear to separate at a horizon. Simply stated, without the observer of that world, there would be no observable world.

What if we use the many world interpretation for our system of entangled spin variables? There is no quantum state reduction. Every possible observable state is an actual state that is actually observed by an observer. An observer is present for every possibility. No signal is transmitted across the universe since no quantum state reduction is performed. A quantum state reduction must disturb the universe as the universe is measured. A measurement disturbs the universe since the universe is not the same after a measurement is performed. A measurement is a quantum state reduction that reduces the quantum state of potentiality for the universe to an actual state, which is a choice, and in that process of choosing, some of that potentiality is lost. With the many world interpretation, there is no quantum state reduction, there is no choice, and there is never any loss of potentiality.

With the many world interpretation, the universe is never disturbed. The natural way to understand the many world interpretation is with the holographic principle. Any possible state of information for the universe is defined on an event horizon, and is observed by an observer present at the central point of view. The event horizon acts like a holographic viewing screen, and defines a state of information. Whatever is observed in that world is like the projection of a holographic image to the central point of view. As that state of information for the world arises, an observer arises at that focal point of perception.

The results for the measurement of entangled spin variables by experimenters that are physically separated from each other violates Bell's theorem (Shimony 2009, 11). These results also seem to violate causality, since the results are instantaneously determined in physically separate locations. The problem with Bell's theorem is it assumes the results of two physically separate measurements are independent of each other, and that the two particles behave as independent entities after they become physically separated. Quantum entanglement directly refutes this assumption of independent behavior.

Bell's theorem assumes the independent behavior of physically separated particles, but also assumes the independent existence of different observers in the same world with the particles. Those two observers observe the observable values of the particles. There is an implicit assumptions the two observers exist within the same world with the particles.

There are two important assumptions lurking in Bell's theorem:
1. The independent existence of multiple particles within the world.
2. The existence of multiple observers within the same world with multiple particles.

The holographic principle resolves this paradox in a very straightforward way, since it demonstrates that neither of these assumptions is correct. In a holographic world, all the information for that world is encoded on the surface of an event horizon, which acts as a holographic viewing screen. Each pixel on the screen encodes a bit of information. The event horizon is a spherical surface that is as far out in space as the observer present at the central point of view of that sphere can see things in space due to the constancy of the speed of light. An observer that is not in a state of free fall through empty space, which is to say that observer is in an accelerating frame of reference, observes an event horizon, where all the information for that world is holographically encoded.

The event horizon always arises from the point of view of the observer present at the central point of view of that sphere. Every observable value observed in that world is like the projection of an image from the viewing screen to that focal point of perception. All the information for the image is coherently encoded on the screen. A measurement of the spin value of a particle in that world is like the holographic projection of an observable image from the viewing screen to that point of view.

A different observer located at a different point of view observes its own event horizon, which also acts as a holographic viewing screen that encodes bits of information. Each viewing screen defines a state of information. The information encoded on one viewing screen is correlated with the information encoded on the other viewing screen due to quantum entanglement. A measurement of the spin value of the second particle by the second observer is always correlated with the measurement of the spin value of the first particle by the first observer due to the quantum entanglement of bits of information on the different viewing screens. Those states of information are entangled.

The two observers do not exist within the same world. Each observer has its own world defined on its own viewing screen. What appears to happen in either world is only correlated with what appears to happen in the other world due to quantum entanglement of bits of information on different viewing screens that define those different worlds, each of which is observed from a different point of view. A viewing screen defines a state of information that is entangled with other states of information.

Quantum entanglement is the nature of coherent organization of information. Coherent organization arises as bits of information tend to align together. Spin networks explain how this is possible. Spin angular momentum tends to combine together due to entangled spin states. As spin combines together, the spin variables tend to align together like little magnets. Alignment of information allows information to become coherently organized on a viewing screen, and makes the viewing screen holographic in nature. Alignment of information also allows for correlation of information between different viewing screens.

The nature of a spin network follows from the above example of entangled spin variables (Penrose 2005, 32.6). A simple spin network is diagramed by Penrose in figure 32.10 of *The Road to Reality*, and needs to be consulted in order to continue this discussion.

With the idea of a spin network, we can see that every event is a decision point where the quantum state branches into alternative paths. This is exactly what Penrose diagrams in figure 32.10. As two spin 0 variables decay into two entangled spin ½ variables, and two spin ½ variables combine together, another decision point is formed, and another branch forms. Two spin ½ variables can combine into a spin 0 or spin 1 variable, but spin 1 is more likely than spin 0 (with probability ¾ versus ¼). The difference in probability arises from the different number of spin eigenstates (one for spin 0, two for spin ½, and three for spin 1). A spin 0 variable only has the spin eigenstate of 0, a spin ½ variable has the eigenstates of +½ and −½, and a spin 1 variable has the eigenstates of +1, 0, and −1.

As entangled spin states combine together, more decision points form, and more branches form. If an observer is present for any event, then that observer measures the amount of spin in that branch. The key ideas are that every decision point is an event where the path branches, and that an observer may be present for an event to measure the amount of spin in that branch. If an observer is present for every decision point, and for every branch, there is no need for the idea of quantum state reduction, and the many world interpretation is the natural way to understand measurement.

In the many world sense, a particular path defines its own world, and an observer is always present for every decision point on that path, or for every event in that world. An observer is present at every decision point, and measures the amount of spin in that path for that event. The only way we can recover the standard interpretation, with the idea of a quantum state reduction, is if an observer is only present for one particular path. The observer still measures the amount of spin on the path at a decision point, but an observer need not be present for every event, and need not be present for every path. The only real difference between the many world and standard interpretations is whether an observer is present at every decision point or not. This obviously has something to do with the nature of consciousness, observation and measurement, or what is actually observed to happen, but not with the nature of physics. The physics of a spin network is exactly the same in either case, since the quantum state of potentiality of a spin network is described by all possible paths that can be drawn in the manner of the diagram of figure 32.10.

The key idea is that an event is a decision point where the path branches. The quantum state of potentiality includes all possible paths, which is the same as all possible diagrams. An actual measurement of spin on any particular path at any particular decision point is observed by an observer for that particular event. In the sense of the standard interpretation, that measurement is a quantum state reduction that alters the quantum state of potentiality by truncating all paths that are not consistent with that measurement. That measurement is like a resetting of initial conditions, and so alters the quantum state. Only the paths consistent with that measurement, or with that resetting of initial conditions, remain in the truncated quantum state of potentiality.

The quantum state of potentiality, which is like a probability distribution, has changed as a consequence of that measurement, but only in the sense of quantum state reduction. Subsequent measurements are conditional on that measurement, since the truncated probability distribution has changed as a consequence of that measurement. The nature of conditional probability, usually stated as Bayes' theorem, has something to do with the memory of events based on observation of events, in the sense all possible subsequent events are dependent on resetting of initial conditions, as occurs with a measurement.

The important point about the standard interpretation is that the quantum state of potentiality is only truncated as a consequence of a measurement. There are no new paths created as a consequence of the measurement, only truncation of those paths that are not consistent with the measurement. A quantum state reduction is only a truncation of some possible paths. With the many world interpretation, there is no truncation of the quantum state. There is no quantum state reduction. Every decision point is an event where the path branches, and all possible paths are included in the quantum state of potentiality. An observer is present for every event to measure the amount of spin on that path for that event. If an observer is present for every event, then there is no need for any quantum state reduction. As far as that observer is concerned, there is no way to distinguish this state of affairs from a quantum state reduction, since the physics does not change.

There is no possible way for an observer to distinguish many worlds from a single world, since all a quantum state reduction does is truncate some possible paths. In the sense of an interference pattern, the paths that are truncated do not interfere with the observed path. As far as any observer is concerned, there is only one world, which is the observer's own world, within which measurement, or quantum state reduction, appears to take place. A quantum state reduction is referred to as the collapse of the wave function precisely because of the destruction of the interference pattern (Davies 1977, 172).

Only those worlds that become entangled, and interfere with each other, can share information. Different worlds that do not become entangled, and do not interfere with each other, do not share information. Those non-interfering worlds only exist in the sense of an ensemble of worlds, which is an essential aspect of any thermodynamic description of the world, as in inflationary cosmology. That ensemble of worlds describes all possible ways in which the universe can initially be created and evolve over the course of time, as energy flows through the universe in the sense of thermodynamics (Davies 1977, 172).

The only way to understand how this state of affairs is possible is if the consciousness of the observer in some sense is 'external' to the world that is measured, and if the laws of quantum theory do not apply to the nature of consciousness (Davies 1977, 172). That is exactly what the holographic principle demonstrates. As a world holographically arises on a viewing screen, the nature of consciousness arises at a point of view. That particular observer is only aware of its own world, and is not aware of the other non-interfering worlds. In the sense of the branching of the quantum state of potentiality, those different world are parallel to each other, but do not interfere with each other. In the sense of inflationary cosmology, a cosmic event horizon is like a bubble in the void, which always has an observer present at the central point of view, and at the center of its own world.

In the sense of inflationary cosmology, each cosmic event horizon defines its own world with an observer present at the central point of view. The event horizon encodes bits of information, and acts like a holographic viewing screen. Every viewing screen defines a state of information, with one bit of information encoded per pixel on the screen. A particular event horizon is observed from a particular point of view. With every event, the quantum state that describes a particular event horizon branches. Every event is a decision point where the quantum state branches due to all the different ways in which information can become encoded on all the different pixels. Those different branches of the quantum state for any particular event horizon do not interfere with each other, and define non-interfering parallel worlds. The quantum state of potentiality for the universe includes all possible event horizons, observed from all possible points of view. A state of information for one particular event horizon may interfere with a state of information for another event horizon, as those states of information become entangled. The observer at the central point of view of any particular event horizon observes its own world, as the path of that world branches away from the path of parallel worlds. Every event is a decision point where the quantum state branches. The most likely path, in the sense of quantum probability, is the path of least action. If the path of least action is always taken, then no decision is ever made. That is the natural way for its world to evolve over time.

Inflationary cosmology tells us that the flow of energy through the universe begins with a big bang event, and flows in the sense of thermodynamics (Penrose 2005, 27.7), as the universe expands in size from the big bang event. That expansion will finally end with the heat death of the universe. If the universe expands in size indefinitely, its temperature approaches absolute zero as its size approaches infinity. In an exponentially expanding universe with dark energy, all other matter and energy will eventually cross the cosmic event horizon, and that universe will only contain dark energy (Susskind 2008, 437). The universe suffers heat death as its temperature approaches the temperature of the maximally inflated cosmic event horizon. The flow of energy tends to flow from a hotter to a colder object. The universe was hottest at the time of the big bang event, since the cosmic event horizon was smallest at that time (Greene 1999, 356), and its eventual heat death occurs with the largest possible cosmic event horizon.

The universe inflates in size due to an unstable process that alters the amount of dark energy, which increases the distance to the cosmic event horizon. As the universe expands it cools, and undergoes a phase transition. The big bang is the state of highest gravitational potential energy, conceptualized as a nearly uniform distribution of matter and energy in space-time geometry (Penrose 2005, 27.11). In inflationary cosmology, the total energy of the universe is zero (Penrose 2005, 28.10), since the negative potential energy of gravitational attraction cancels out all forms of positive energy. A total energy of zero is like an energy that determines an escape velocity (Penrose 2005, 27.11). Matter and energy fall together from the big bang under the influence of gravitational attraction. As the universe expands in size it also cools, and it changes phase. That change in phase is like the burning that occurs as an unstable state of high potential energy transitions to a more stable state, and releases heat. Heat flows from the hotter to colder object, and nothing is colder than the maximally inflated cosmic event horizon (Bousso 2002, 44).

Energy flows through the universe in the sense of thermodynamics. Things move around due to kinetic energy, and tend to scatter off each other in collisions as they exchange some kind of radiation, which is the nature of a force (Susskind 2008, 162, 328, 346). In quantum field theory, a force is conceptualized as an exchange of particles between other particles. As those particles move around, they tend to scatter in random directions. We measure kinetic energy as heat at the macroscopic level when motion becomes randomly directed at the microscopic level. Things are hotter at the macroscopic level if there is more disordered kinetic energy at the microscopic level. Heat tends to flow from hotter to colder objects since hotter objects radiate away more heat. Heat is some form of radiation that carries away disordered kinetic energy. Things tend to scatter off each other in collisions, but also tend to bind together due to their potential energy of attraction. Bound states form at the microscopic level, but they also form at the macroscopic level, which we call a phase transition. Formation of a bound state alters the balance between kinetic and potential energy. Kinetic energy is radiated away as any bound state forms, just as heat is radiated away from liquid water as ice forms. The formation of a bound state is always like a scattering event with something else in the universe due to that radiation of energy. Only that flow of energy allows form to become transformed into new form.

Bound states form through a process of symmetry breaking, which leads to the formation of a more stable equilibrium state through a reduction in the amount of symmetry

(Greene 1999, 351). Symmetry breaking involves the alignment of bits of information, just like little magnets that tend to align together (Penrose 2005, 28.1). In the usual physical systems that undergo phase transitions, like liquid water that freezes into ice, the balance between kinetic and potential energy is shifted in favor of potential energy as disordered kinetic energy is radiated away. But that balance can also be shifted if potential energy is added to the system, as occurs when a biological organism adds the high potential energy of a biological molecule to its body through a process of eating, burns that molecule within its body, and excretes away the disordered kinetic energy. All body growth and development requires some kind of biological symmetry breaking, as does the maintenance of body stability (Damasio 1999, 138). Self-replication of the form of a body is only possible within a coherently organized phase of organization. That coherent organization only develops through a process of symmetry breaking.

There are two mysteries of the physical world that science has a difficult time to explain. The first mystery has to do with the computational nature of information. All theories of the physical world are computational in nature, and are expressed as equations that relate one physical value to another physical value. A physical value refers to an observable property of some physical thing that can be measured, such as the mass, electric charge, or spin angular momentum of an electron. If we think of the electron as a point particle, its location in space and time are also measurable values. At the most fundamental level possible, those measurable values can be reduced to bits of information, which John Wheeler famously expressed as 'It from bit' (Susskind 2008, 136). In the sense of computer processing, the laws of physics are only computational rules that describe how those bits of information are dynamically processed over a sequence of events.

If we think of the physical universe as a computer that processes information, the laws of physics describe how bits of information are dynamically processed over a sequence of events (Susskind 2008, 137). Every event is like a processing cycle that updates the configuration state of bits of information from one moment to the next moment. The laws of physics, as typically expressed in terms of equations, are like computer programs, or computational rules, that underlie this processing of information.

The first mystery of the physical world is about the origin of the information and the origin of the programs that dynamically process that information. If we equate a program with an equation, like Maxwell's equations, the Dirac equation or Einstein's equations, we have to explain where both the information and the programs come from. The beauty of a unified theory, like string theory, is all of these equations naturally arise from fundamental symmetry principles that represent the symmetry of empty space.

The most basic symmetry of empty space is general coordinate invariance (Zee 2003, 76), which is usually expressed as the principle of equivalence. Einstein's field equations for the metric naturally arise from general coordinate invariance (Zee 2003, 419). String theory is based on the fact that the only coordinate invariant 'area' is of the world sheet of a string (Zee 2003, 452), just as the only coordinate invariant 'length' is of the world line of a point particle. General coordinate invariance is the fundamental gauge invariance.

The U(1), SU(2), and SU(3) gauge symmetries of particle physics naturally arise with compactification of extra dimensions (Zee 2003, 433). Even the Dirac equation arises naturally if we incorporate super-symmetry into the symmetry of empty space. In this sense, the computer programs that describe the dynamical processing of information naturally arise from the symmetry of empty space. The encoding of information also naturally arises from symmetry, since a compactified dimension encodes bits of information like a spin network, or in the sense of loop variables (Penrose 2005, 32.4).

The second mystery of the physical world is the second law of thermodynamics, and the origin of order. The second law describes irreversibility, as energy flows from a more ordered to a less ordered state. The only natural way the universe evolves over time from a more ordered state to a less ordered state is if the initial state of the universe is ordered (Feynman 1963, I 46-7). But if the universe begins with a big bang event that is only a quantum fluctuation, there is no natural way to put that order in the initial state of the universe. Only with all possible initial states of the universe (in the sense of an ensemble of universes, or many worlds) will some of those initial states naturally become ordered.

The only natural way to have irreversibility is if the initial state of the universe is ordered. How does order arise in the initial state of the universe? If the universe only contained thermal blackbody radiation, it would rapidly come into thermal equilibrium (Penrose 2005, 27.13). Even as the universe inflates in size, if it only contained thermal blackbody radiation, it would rapidly come into thermal equilibrium at a lower temperature as the cosmic event horizon inflates in size. In the sense of inflationary cosmology, the big bang event is only a quantum fluctuation, and the most likely way for that fluctuation to occur is for the universe to only contain thermal blackbody radiation (Zee 2003, 403).

But the universe does not just contain thermal blackbody radiation. The initial state of the universe is conceptualized as a nearly uniform distribution of matter and energy (Penrose 2005, 27.11). Most of that matter is initially in the form of a very hot gas of protons and electrons that are overall electrically neutral (Greene 1999, 346). As the universe expands and cools, the electrons bind to the protons to form hydrogen atoms. The hydrogen atoms gravitationally attract each other and tend to clump together into nebulae, and then into stars, galaxies and planets. Protons within stars tend to fuse together into larger atomic nuclei. As protons fuse together deep within stars, very high energy photons are released, which gives rise to the electromagnetic radiation that is radiated away from the star. The radiation of energy away from the hot surface of stars into colder outer space indicates the universe is not in thermodynamic equilibrium. That radiation of energy reflects the order inherent in the initial state of the universe, which arises from a nearly uniform distribution of protons and electrons early in the history of the universe.

The formation of stars that arise as hydrogen atoms tend to gravitationally clump together indicates the universe is not in thermodynamic equilibrium. The formation of stars, and the fusion of protons together deep within stars to form atomic nuclei, reflects that state of thermodynamic disequilibrium, as heat flows from hotter to colder objects, energy flows from more ordered to less ordered states, and total entropy increases. That overall increase in entropy for the universe only arises from the initial ordered state of the universe, with a nearly uniform distribution of protons and electrons.

Total entropy increases as protons fuse together deep within stars (Davies 1977, 92). The most significant source of increased entropy in the local environment of a star is the fusion of protons within the star. The temperature deep within a star is millions of degrees, while the temperature at the star's surface is only a few thousands of degrees. As protons fuse together within a star, the process of fusion releases high energy photons. A photon must be radiated away from the protons for the velocity of the protons to fall below an escape velocity, as the potential energy of nuclear attraction between the protons overcomes their kinetic energy. The fusion of protons into larger atomic nuclei is a kind of nuclear 'burning' of protons. Deep within a star electrons are ionized due to the very high temperature, and matter is in the form of an ionized plasma of electrons and atomic nuclei. A high energy photon released as protons fuse together has an energy in the range of millions of electron volts, which reflects the mass energy of fusion by $E=mc^2$, as about 1% of the mass of a proton is converted into electromagnetic energy. As a high energy photon is released into the ionized plasma, it tends to scatter off the negatively charged electrons and positively charged atomic nuclei. With each scattering event there is a tendency for a high energy photon to become dispersed into lower energy photons. More and more dispersion occurs as those photons make their way out to the cooler surface of the star. Those dispersed photons have a lower energy and a lower frequency as $E=h\nu$. The photons that are eventually radiated away from the star only have an energy in the range of an electron volt. The dispersion of a high energy photon into millions of lower energy photons radiated away into space is a huge increase in entropy. But the thermodynamic disequilibrium of a star, with a very hot interior of the star, a cooler surface of the star, and cold outer space, only arises from gravitational collapse.

The initial order of the universe is inherent in the gravitational potential energy of a nearly uniform distribution of protons and electrons early in the history of the universe (Penrose 2005, figure 27.10). The difficult thing to explain is how that initial order arises if the big bang event is only a quantum fluctuation. Inflationary cosmology postulates an ensemble of universes that include all possible initial states. A thermodynamic ensemble will naturally include some initial states that are highly ordered (Davies 1977, 199).

The total entropy of a star and its environment increases as the star radiates away photons into cold outer space. As the star burns, heat is radiated away into the environment, like any other kind of burning. That heat is disordered kinetic energy. Entropy increases due to the disordered kinetic energy carried away from the star by the photons. The system of the star and its environment is not in thermodynamic equilibrium, since total entropy has not yet become maximal. That maximal state of entropy only occurs if the star collapses into a black hole. The black hole is the system of the star and its environment in a maximally disordered state, with maximal entropy. That maximally disordered state is in thermal equilibrium. Hawking radiation is only a kind of thermal blackbody radiation.

The largest possible increase in entropy occurs if the star collapses into a black hole. As any system forms a bound state (like liquid water that freezes into ice), the entropy of that system appears to decrease locally. This is only possible due to heat irreversibly radiated away into the environment as the bound state forms. The overall entropy of system plus environment increases due to the heat radiated away. As any system collapses into a black hole, the total entropy of the system plus its environment becomes maximal. The entropy of a black hole is proportional to the surface area of its event horizon. The only way to understand entropy is if all the information for the system and its environment are defined on the surface of an event horizon, which is the holographic principle.

The holographic principle tells us that whatever forms we observe in space is like a holographic projection of images from a surface (the viewing screen) to a focal point of perception (a point of view). All the information for that world is defined on the viewing screen, which defines a state of information that (thermo)-dynamically evolves into other states of information as energy flows from more ordered to less ordered states. The problem is how to put order into the initial state. The only natural solution to this problem is to have an ensemble of initial states, in the sense of many worlds. Otherwise, that order has to be put in by hand, as in the hand of God (Penrose 2005, figure 27.21).

The concept of an observer-centric world is inherent in the covariant entropy bound of the holographic principle (Banks 2011). This principle demonstrates that there is no single objective reality as implied in quantum field theory, but many entangled worlds that share information with each other, each defined on its own holographic screen, and each observed from its own point of view. Every observer observes its own world, with observable values defined in its world by a time-dependent Hamiltonian operator and Hilbert space (Bousso 2002, 42; Banks 2011). The world as observed from a point of view in the Milky Way galaxy is not the same world as observed from a point of view in a galaxy on the other side of the universe. Different observers observe different worlds that are limited by an event horizon, which defines each observer's observable world.

No observer can see beyond its event horizon. The event horizon limits observations in the observer's world. If that event horizon changes with time, as it does in inflationary cosmology, or if matter and energy appear to cross the observer's event horizon and disappear from the observer's world, the observable values of that world also change with time. This implies a time-dependent Hamiltonian and Hilbert space for each observer. The only constraint is that the observable values of one observer must be consistent with those of another observer if the Hilbert spaces overlap and share information.

If two observers share information with each other, and their observable worlds overlap in the sense they share a part of the same Hilbert space, and share observable values together, then the observations of one observer must be consistent with the observations of the other observer. If there is no overlap, and one observer's world is totally beyond the event horizon of the other observer's world, then they do not share any information with each other. The event horizon of each observer defines the observable values that are observable in that observable world. Those observable values can change over time, either if the event horizon changes, in the sense of inflationary cosmology, or if matter and energy appear to cross the event horizon, and become unobservable. The radius of a cosmic event horizon, as determined from a positive cosmological constant, gives the maximal size of such an observable world in inflationary cosmology.

Every observer in an accelerating frame of reference observes its own event horizon. An accelerating frame of reference is like an accelerating trajectory through space, which defines the world-line of an observer. An accelerating frame of reference only arises with expenditure of energy, just like a rocket ship that accelerates through space as it expends energy. The event horizon arises with expenditure of energy. That event horizon appears to radiate energy, which is called the Unruh effect (Smolin 2001, figure 16). Radiation of energy from the event horizon is due to separation of virtual particle-antiparticle pairs at the horizon, just like Hawking radiation from the event horizon of a black hole.

The event horizon always arises from the point of view of the observer, and is as far as the observer can see things in space due to the constancy of the speed of light. If the expenditure of energy comes to an end, that acceleration also comes to an end, the observer enters into a state of free fall through empty space, and the event horizon disappears. A viewing screen is an event horizon that encodes information, and appears to radiate energy, along the lines of the Unruh effect.

Along the lines of the holographic principle, all the information for the observer's observable world is encoded on its holographic viewing screen. This is explicitly demonstrated in string theory, but only for the special case of anti-de Sitter space with a negative cosmological constant. The world we actually appear to live in is an exponentially expanding world with a positive cosmological constant, which is described in relativity theory as de Sitter space (Bousso 2002, 42; Banks 2011).

Every observer in an accelerating frame of reference observes a viewing screen from its own point of view, which shares information with other viewing screens observed from other points of view, but only to the degree those viewing screens are entangled with each other, and those observable spaces overlap. This is the case in de Sitter space with a positive cosmological constant. The problem of how different viewing screens encode information in de Sitter space is more complicated than in anti-de Sitter space, since each observer has its own time-dependent Hamiltonian and Hilbert space of observable values, which change over time as the observable world that is limited by the observer's event horizon changes over time. The problem is much simpler in anti-de Sitter space, since there is only one viewing screen (the boundary of anti-deSitter space) and one observer at the central point of view (Bousso 2002, 42). The boundary of anti-de Sitter space has a Killing vector, like the event horizon of a black hole, and there is a sense in which energy is conserved that allows for a single time-independent Hamiltonian and Hilbert space of observable values (Banks 2011). The event horizon of an observer in an accelerating frame of reference in de Sitter space does not have a Killing vector, and so there is no sense in de Sitter space of the conservation of energy, or a time-independent Hamiltonian and Hilbert space of observable values. That does not mean that the holographic principle does not apply in de Sitter space, only that its formulation is a much tougher nut to crack.

String theory is inherently holographic, but there is no unique formulation of string theory, and no one master set of equations (Susskind 2008, 309). String theory can be formulated in many different ways in terms of different geometrical objects, referred to as D-branes. One formulation tends to morph into another formulation. D-branes represent mathematical objects, like points, strings, and membranes, but none of these geometrical objects are fundamental in nature. They are all holographic in nature. This is exactly the same problem with the current quantum field formulation of inflationary cosmology.

Inflationary cosmology is based upon the idea of vacuum energy, which in the sense of relativity theory is equivalent to a non-zero cosmological constant. The idea of a bubble universe arises from spontaneous symmetry breaking, in the sense of a tunneling event from a vacua of higher vacuum energy to a vacua of lower vacuum energy. In the sense of quantum field theory, the nature of vacuum energy is understood to arise from a special kind of Higgs field, referred to as an inflaton field. The amount of field energy is assumed to depend on the field amplitude in a complex way, with a potential energy curve that has multiple local minimum values, which specify the amount of vacuum energy for each possible vacua. A bubble universe can inflate in size if there is a quantum tunneling event from a higher energy to a lower energy vacua (Green 2011, figure 6.5).

The problem with the quantum field theory formulation of inflationary cosmology is that it becomes invalid at the Planck level (Banks 2011). The whole concept of vacuum energy is flawed in quantum field theory. Vacuum energy arises from a Higgs field through spontaneous symmetry breaking, which requires tunneling events between false vacua. The Higgs field is a field theory concept, which only gives a holographic description of events, and describes vacuum energy as filling space. The more fundamental description is the viewing screen description. In the viewing screen description, empty space is truly empty. All excited states of energy are defined on the viewing screen. This is as much the case for vacuum energy as for any kind of energy.

The quantum field formulation of inflationary cosmology is a flawed description, which is corrected with a holographic viewing screen description (Banks 2011). The mathematical details are still obscure, but the physical principles are easy to understand, just as Einstein used the principle of equivalence to intuitively understand relativity theory before he discovered field equations for the metric.

The quantum field description of vacuum energy, with a pre-existing Higgs field, and virtual creation of particle-antiparticle pairs in empty space, is a holographic description. All particle excitations of field energy, whether real or virtual, are holographic in nature. In the viewing screen description, empty space is truly empty. All excited states of energy and information (excited relative to the ground state of zero energy and no information of empty space) are defined on a viewing screen. A state of non-zero but locally minimum vacuum energy is also an excited state.

A non-zero cosmological constant in quantum field theory implies symmetry breaking and an excited state. In the sense of holography, all excited states are defined on surfaces of quantized space-time, which act as holographic viewing screens that encode information, as observed from the central point of view of an observer. With the viewing screen description, empty space is truly empty. Empty space embodies unbroken super-symmetry, which allows for a zero cosmological constant (Banks 2011). Vacuum energy is exactly zero if empty space exactly embodies super-symmetry. Only with symmetry breaking can a non-zero cosmological constant arise, which specifies an excited state.

It appears the vacuum state is characterized by alternative vacua, with minimum but non-zero vacuum energy, but that is only a holographic description. In terms of a quantum field, vacuum energy can take on a minimum but non-zero energy, but that is only a holographic description of what appears to happen in empty space. The virtual particle-antiparticle pairs that give rise to vacuum energy are as holographic as anything else that appears to happen in empty space. In terms of the viewing screen description, all possible excited states of information and energy are defined on a viewing screen. The alternative vacuum states of a quantum field theory are also defined on viewing screens in terms of alternative states of information and energy. In the viewing screen description, empty space is truly empty. It is the ground state of zero energy from which all excited states arise. Those excited states include states classified as alternative vacuum states in the quantum field description. A non-zero vacuum energy is characteristic of excited states.

The other problem with the quantum field description is the idea of a single objective reality of matter and energy that arise within some pre-existing space and time. In this kind of an objective reality, an observer can only arise from the behavior of matter and energy, but since the observer observes the behavior of matter and energy, there is a paradox of self-reference. There is no paradox of self-reference in a viewing screen description. The observer arises at a point of view while a world arises on a viewing screen. There is no single objective reality, but many entangled worlds that share information with each other, each observed from its own point of view. The viewing screen and the point of view of the observer both arise within empty space.

The mistake quantum field theory makes is the assumption of a single objective reality of matter and energy in some pre-existing space and time. Holography demonstrates that consensual reality is composed of multiple entangled worlds, each defined on its own viewing screen, and each observed from its own point of view. Both a world defined on a viewing screen and an observer present at a point of view arise from, and within, empty space. The expenditure of energy that animates the images of that world, as displayed on the viewing screen, is inherent in the accelerating frame of reference of the observer.

So far the holographic principle has only been used to explain the connection between unified theories, like string theory, and more conventional theories, like relativity theory and quantum field theories. The more fundamental description of a unified theory is the viewing screen description, and our more conventional theories only give a holographic description of events in the world. A quantum field theory is inherently a description of the behavior of particles, like an electron or a photon, in 3+1 dimensional space-time. That holographic description of events always has a corresponding viewing screen description, where all the fundamental bits of information are defined. The observed behavior of a particle is like the holographic projection of an image from the viewing screen to a point of view, as those images are animated over a sequence of events.

We are now about to make the leap from the behavior of particles to the behavior of bodies in the world. This is not a small conceptual leap. The number of protons in the observable universe is approximately 10^{80}, which corresponds to about 10^{123} fundamental bits of information, based on the amount of information encoded in a black hole with the same mass (Penrose 2005, 27.13). A similar number arises from the size of the cosmic event horizon at 15 billion light years. A biological body with a mass of 100 kg has about 10^{29} protons. How can we generalize from the relatively simple behavior of particles to the complex behavior of biological bodies? The answer is coherent organization, which is as valid for the behavior of a biological body as for the behavior of a particle. The form of every observable thing observed in the world is a coherently organized bound state of information. This is as much the case for a particle as for a biological body.

It is worth noting the evolutionary concept of the survival of the fittest body follows directly from the concepts of symmetry breaking and quantum probability. A body is a bound state of information that only develops through a process of biological symmetry breaking. The nature of biological symmetry breaking is the balance between potential and kinetic energy in any biological organism is altered in favor of potential energy as potential energy is added to the organism and disordered kinetic energy is radiated away. The addition of potential energy to the organism is what we call eating. All biological development, growth, behavior and survival requires a process of eating, as high potential energy molecules are added to the organism, burned within that body, and disordered kinetic energy is excreted. A body only develops and survives through a process of eating, which is an aspect of biological symmetry breaking. The bodies most likely to survive, in the sense of the survival of the fittest, follows from quantum probability. The most likely path is the path of least action. The path of least action minimizes the expenditure of kinetic energy while maximizing the preservation of potential energy.

As energy flows in its universal gradient from the big bang event to the heat death of the universe, those bodies that most efficiently transfer energy down this universal gradient are most likely to survive, since they follow the path of least action. The flow of time arises as energy flows in its universal gradient. It is also worth noting the principle of equivalence, not quantum theory, describes the nature of consciousness. Quantum theory specifies that every observable value of the quantum state is observed by an observer, but has nothing to say about the nature of the observer. The principle of equivalence specifies an observer arises at a point of view in empty space as a world arises on an event horizon.

We are now ready for another conceptual leap. What is the nature of a mind? Does a mind arise from a body, or is a body somehow dependent on a mind? Neuroscience implicitly assumes that a mind arises from a body, but that assumption has never been verified. The problem with this assumption is that it assumes matter and energy exist within some pre-existing space and time. The body is assumed to be composed of matter and energy that exist within space and time, and somehow the mind is assumed to arise from the body. We know from the holographic principle that this assumption is incorrect. The other problem with this assumption is that it is too limited to explain the mind.

It is impossible to explain the nature of the mind with this assumption. This assumption is a mental model of the world, or a mental concept that arises in a mind. The content of the mind is information content. Behavior and emotional expression arises with the flow of energy through the world. This mental concept of the world can never explain the nature of consciousness that perceives information and energy in that world. The mind displays an entire world that includes the body. How can the mind arise within that world? How can consciousness arise in the same world the body arises within, since consciousness perceives the entire world that the mind displays? That world includes the body.

A mind displays an entire world that includes the central form of a body. The brain is part of that body. The entire world the mind displays is perceived by a presence of consciousness. This state of affairs is apparent for everyone to see, but what is so odd is that this state of affairs is so universally ignored in the scientific world. The problem with the assumption that consciousness arises inside a brain is its logical inconsistency, which contradicts the mathematical consistency of the holographic principle. The nature of the mind can only be understood in the sense of a viewing screen that displays forms of information that arise in a perceivable world, which is perceived from the point of view of a focal point of perception. The viewing screen is only a state of information defined on an event horizon, which always arises from the central point of view of the observer. The principle of equivalence tells us that the event horizon arises from the point of view of the observer present at the center of an accelerating frame of reference. The event horizon is only a two dimensional spherical surface that arises in an accelerating frame of reference, which is as far as the observer present at the central point of view can see things in space due to the constancy of the speed of light. All the information for all the things that are perceived in that space is defined on the surface of the event horizon.

As the event horizon arises in empty space, the observer arises at a point of view. All the information in the brain of a person is defined on the viewing screen. Consciousness cannot arise inside a brain. The brain is defined on the viewing screen, like everything else in that world. Consciousness is outside, present at a point of view in empty space.

The mind is defined on a viewing screen that displays an entire world. The viewing screen is defined on an event horizon that arises from the point of view of the observer. The mind displays an entire world that includes the body of that person, with a brain inside that body. The consciousness that is present for that mind is present at a point of view. The viewing screen, and all the images for that world, only arises if the observer is in an accelerating frame of reference, which is the same as the expenditure of energy. Without that expenditure of energy, there is no accelerating frame of reference, and there is no event horizon. Without that expenditure of energy, the observer enters into a state of free fall through empty space, and that world disappears.

The idea of an observer-centric world can be taken to its logical conclusion. The controversial aspect of this idea is that the observer is no longer identified with the nature of anything the observer can observe in that world. For the purpose of scientific hypothesis, let's assume the observer exists as pure consciousness. The first thing is to define an observer-centric world. The equivalence principle tells us the observer is present at a point of view that follows a world-line in space. If that world-line defines an accelerating frame of reference, that observer always observes an event horizon, which is as far as that observer can see things in space due to the constancy of the speed of light. Everything beyond the event horizon is hidden from the observer.

In an observer-centric world, the viewing screen defines a state of information, with one bit of information encoded per pixel on the screen. That state of information defines an entire world that only arises from the point of view of the observer of that world. The viewing screen displays an entire world. That world only appears from the point of view of the observer. That state of information is defined by the way bits of information are encoded on the pixels. In the sense of quantum theory, every event is a decision point where the quantum state of that world branches, due to all the different ways in which bits of information can become encoded on all the pixels of the viewing screen. Coherent organization of information allows for the development of observable forms of information, which self-replicate in form over a sequence of events, and are the nature of the observable images projected from the viewing screen to the observer at the central point of view. In the sense of an animation of images, the behaviors of those self-replicating forms are enacted over a sequence of events in the flow of energy. Thermodynamics allows us to understand the nature of that flow of energy (and the flow of time), in the sense that energy tends to flow from more ordered to less ordered states.

Every observer observes its own world from its own point of view. That world is nothing more than forms animated on a viewing screen that appear three dimensional since the forms are holographic. Forms tend to self-replicate in form over a sequence of events in the flow of energy due to coherent organization, as an animation of forms is animated.

The principle of equivalence helps us understand how an animation is animated. As the observer focuses its attention on a world of form, there is an expenditure of energy. That expenditure of energy puts the observer in an accelerating frame of reference, just like a rocket ship that expends energy as it accelerates through empty space. The difficult thing to wrap our minds around is there is no such thing as a rocket ship, except as an observable image that is projected from a viewing screen to a point of view. An event horizon always appears from that accelerated point of view. Every observer that expends energy is placed in an accelerating frame of reference, and observes an event horizon, which holographically defines the entire world of form that the observer observes.

The observer's world of form always appears from its own point of view, as it focuses its attention on those forms. There is an expenditure of energy, which places the observer in an accelerating frame of reference, and leads to the animation of that world. As the observer focuses its attention upon those forms, there is an investment of emotional energy in that animation. That investment of emotional energy animates the form of the observer's body, which is the central form of its world. The viewing screen displays an entire world, but the central form of a particular body is always displayed on the viewing screen from the central point of view of that particular observer.

The observer's world, perceived from its point of view, shares information with other worlds, perceived from other points of view, since those different worlds are entangled. Each world is defined on a viewing screen that defines a state of information, but different states of information can become entangled with each other and share information, and so different forms can appear on each viewing screen in addition to the central form. The observer's investment of emotional energy animates the form of its own body, which is the central form of its world.

The investment of emotional energy from another point of view animates the central form of another world. Other forms can appear in any observer's world since those different worlds are entangled and share information. The focus of attention of the observer of any particular world from any particular point of view leads to an investment of emotional energy in that world, which animates the central form of a body in that world. Other bodies are animated in that world due to the focus of attention of the other observers, each of which observes their own world from their own point of view. Those different worlds are animated together since those different states of information are entangled together. Collectively, those entangled states of information define consensual reality, which is not a single objective reality, but many different worlds that are each observed from their own point of view, and which only share information with each other.

From the point of view of any particular observer, the observer's entire world is displayed on a viewing screen that defines a state of information. That world is animated as energy flows in the sense of thermodynamics. A state of information is defined by the way information is encoded on all the pixels on the screen. In the sense of quantum theory, every event is a decision point where the quantum state branches, since information can become encoded in many different ways. The only reason we have a sense of the flow of time is due to the second law, as states of information tend to become more disordered. That increase in entropy applies to the entire world displayed on the viewing screen. Within that world, a form of information may become more ordered, and may coherently self-replicate its form, as long as the entire world becomes more disordered. A local increase in order of a particular form can only occur at the expense of some other form, which becomes more disordered. This is obviously a problem if every observer wants to survive in the form of its own body. Emotional expressions inherently are self-defensive in nature, since they defend the survival, and self-replication, of the form of a particular body observed from a particular point of view.

In spite of this problem, there is a natural way for the universe to evolve over time. That natural evolution takes the path of least action, which is the most likely path in the sense of quantum probability, and the most energy efficient way for the universe to evolve. But from the point of view of a particular observer, the path of least action may not maximize the probability of survival of its own body, and so for selfish reasons, an alternative path may be taken in order to maximize the chances of body survival. Although the quantum state constantly branches, the observer of any world only observes a particular path taken in that world, but that path shares information with the path of other entangled worlds.

Even the best neuroscientists in the world, like Antonio Damasio, are confused about the nature of consciousness. Damasio implicitly assumes the world is composed of matter and energy that exist within space and time when he describes the problem of consciousness: "The neurobiology of consciousness faces two problems: the problem of how the movie-in-the-brain is generated, and the problem of how the brain also generates the sense that there is an owner and observer for that movie. The two problems are so intimately related that the latter is nested within the former. In effect, the second problem is that of generating the *appearance* of an owner and observer for the movie *within the movie*" (Damasio 1999, 11). Consciousness is only an *appearance* for Damasio. This assumption places consciousness *within* the same world matter and energy appear to exist within, which is a paradox of self-reference, and which makes that description logically inconsistent. The observer does not arise within the same world with matter and energy.

The holographic principle explains what Damasio calls the 'movie in the mind', except the mind is only a movie of images animated upon a viewing screen. The mind is the viewing screen that displays an entire world, like a bubble in the void. That viewing screen always has an observer present at the central point of view. The confusing aspect of the world is its holographic nature. Organs of sensory perception in a body appear to relay information about an entire world to a brain. That apparent relay of information includes external sensory perceptions of the world and internal perceptions of emotional body feelings, but all of those perceptions are only a holographic appearance. All the fundamental bits of information for the world are defined on the surface of an event horizon, which defines a state of information for an entire world that includes the body. The mind displays an entire world that includes the body. The form of a body is an image on a viewing screen. The form of the body is the central image, and all external sensory perceptions of the world are relayed through that central image, in the same way that all internal emotional perceptions of the body, or body feelings, are relayed through the central image. Consciousness perceives the entire world that the mind displays.

The consciousness of the viewer of that viewing screen is never defined by the information encoded on that viewing screen. That is what the incompleteness theorems prove. The viewer is always outside the viewing screen, present at a point of view. Consciousness can never be reduced to the way information is encoded or coherently organized on any viewing screen it observes. Its true nature is always outside the viewing screen. That viewing screen is only an event horizon that arises within the empty background space of the void, as observed by the observer at that central point of view. As a viewing screen spontaneously arises, consciousness arises at a point of view.

The void is the source of all information and energy. The void is the source of the universe, and everything in the world. The void is a state of zero energy and no information, which physicists call the vacuum state. In this sense, the void is the 'stateless state', since all states of information and energy, and all states of the world, are defined on surfaces of quantized space-time. The stateless state is the most stable state, since it is the unchanging ground state. That empty background space is the 'ground of being', in the sense that it is the source of all things that appear to exist in the world. It is the primordial nature of existence. Everything in the world arises from that 'ground'.

This way of understanding the nature of the void, as the empty background space, vacuum state, or ground state from which all excited states of information and energy arise, is not controversial within mainstream theoretical physics. All unified theories, like string theory, and all theories of the creation of the universe, like inflationary cosmology, assume the existence of the void, and understand the nature of the void in this way. The holographic principle helps us understand that the nature of this empty background space is not the same as our usual 3+1 dimensional space-time, which only has a holographic kind of reality. What is controversial is to understand the nature of the void as the source of consciousness that perceives the form of everything in that holographic world. There is no easy way to say this, so we might as well say it as clearly as possible, and appeal to the great tradition of spiritual wisdom to help us understand what it means.

The nature of the void is the 'infinite nothingness' that is the 'one' source of consciousness that perceives everything in any world that holographically arises from the void. As a world arises on a viewing screen, a presence of consciousness arises at that particular point of view. That viewing screen is the nature of the mind. As that mind arises, a presence of consciousness is divided from its true undivided state with the creation of that world. This is what all the great spiritual writings of the world tell us about the nature of creation. It is found in Genesis, the Rig Veda, and the Tao Te Ching (Lao Tsu 1997).

In the beginning God created the heaven and the earth
And the earth was without form and void
And darkness was upon the face of the deep
And the Spirit of God moved upon the face of the waters
And God said 'Let there be light'; and there was light
And God saw the light, that it was good
And God divided the light from the darkness

The Book of Genesis describes how the 'light of consciousness' is divided from the 'one' source of consciousness with the creation of the world. It says 'the light is divided from the darkness' with the creation of the world. That darkness is the void, which is the source of everything in the world. The 'Spirit of God' is the presence of consciousness that is divided from the 'one' source with the creation of the world. That spirit 'moves upon the face of the deep', as all the images for that world are animated on the surface of an event horizon that arises within the void, and are perceived at a focal point of perception. The focus of attention of that presence of consciousness is focused upon those images.

The event horizon acts as a holographic viewing screen, and projects images to the central point of view, where a presence of consciousness is always present at that focal point of perception. That presence of consciousness is divided from the 'one' source of consciousness, which is called the darkness. The true nature of that source is undivided consciousness. As the viewing screen arises, a presence of consciousness arises at that point of view. It all arises from the void. The face of the 'deep' is a surface, just like the surface of the ocean. That ocean is a void of undifferentiated consciousness.

The Rig Veda describes the process of creation as 'that which becomes through the power of heat', just like modern cosmology. It refers to the source of existence as non-existent, oneness and nothingness. 'Darkness was hidden by darkness in the beginning. All that existed then was void and formless.' The Rig Veda also has a bit more to say about the nature of the creator, who may or may not know its true nature. Although the creator may not know what it is, a sage may look across from the existent to the non-existent:

The non-existent was not; the existent was not at that time
An unfathomable abyss
There was neither death nor immortality
There was not distinction of day or night

That One thing, breathless, breathed by its own nature
Apart from it, there was nothing
Darkness was hidden by darkness in the beginning
All that existed then was void and formless

That which becomes, was born through the power of heat
Upon that desire arose in the beginning the first discharge of thought
Sages discovered this link of the existent to the non-existent
Having searched in the heart with wisdom
Their line of vision was extended across

What was below, what was above?
Who knows truly
Whence this creation came into being

He, the first origin of this creation
Whether he formed it all or did not form it
Whose eye controls this world in highest heaven
Surely he knows, or perhaps he knows not

The Tao also tells us that consciousness arises from the void, and is always present for whatever appears to happen in the world, as everything in the world appears to move. Everything in the world arises from the void as consciousness arises from the void, but the void stands alone, and is silent and unchanging:

In the silence and the void
Standing alone and unchanging
Ever present and in motion
I do not know its name
Call it Tao

The Tao tells us that a divided presence of consciousness may return to its true undivided state, but that return is only possible in a state of being desireless. The Tao refers to the mystery of that undivided state as the darkness, or the void:

Ever desireless one can see the mystery
Ever desiring one can see the manifestations
These two spring from the same source but differ in name
This appears as darkness
Darkness within darkness
The gate to all mystery

The Tao tells us that return to that formless state is only possible in a state of being desireless. If any desires are expressed, then the world is animated in the flow of that energy, and animated forms are manifested and appear to move in that world. The desireless state is only possible with the withdrawal of emotional energy from the world. The flow of energy through the world is what animates that world over a sequence of events, and the withdrawal of that energy is the de-animation of that animation.

When we examine anything in the world, the focus of attention of our consciousness is focused in that examination. We focus our attention on the nature of information and energy in that world. Every expression of desire involves the expenditure of energy. The world is defined on a viewing screen by a state of information, and those states are animated over a sequence of events in the flow of energy like the frames of a movie. A presence of consciousness is always at the center of its own world. To paraphrase from the *Lion King*: 'whatever the light of consciousness touches, is its kingdom'.

The focus of attention of consciousness on the world leads to an investment of emotional energy in that world, which always creates an emotional attachment to something in the world. The focus of attention is only withdrawn from the world if that investment of emotional energy is withdrawn. That withdrawal of emotional energy leads to a detached, desireless state. The willingness to let go of those attachments, withdraw that investment of emotional energy, and do nothing, only arises if the futility of everything that can be done in the world is clearly seen. The focus of attention of consciousness is only shifted away from the world onto that nothingness if nothing is done, in the desireless state. Only a detached observer can return to its true undivided, formless state.

Look, it cannot be seen-it is beyond form
Listen, it cannot be heard-it is beyond sound
Grasp, it cannot be held-it is intangible
These three are indefinable
Therefore they are joined in one

Return to that undivided state is called truth realization or enlightenment, and is described as dissolution into nothingness and oneness. Consciousness dissolves back into its source like a drop dissolves into the ocean. Dissolution only occurs if the observer of that world detaches itself from that world, and enters into a state of free fall through empty space. A state of free fall arises without the expression of desire, and that world disappears. That world is defined on a viewing screen, which is an event horizon. The principle of equivalence tells us that in a state of free fall through empty space the effects of all forces disappear, the event horizon disappears, all forms disappear, and that world disappears.

It is as though the void creates a world for its own amusement, and watches with detached interest from its seat in the audience, while images play on a stage. The only problem with this play of consciousness is self-identification with form, which is the problem of the ego. Every viewing screen that arises is observed from its own point of view, which gives the appearance of an entire world, and the appearance of separation, self and other. The void reveals itself to itself through all the things that appear to exist in a world. In that world, self appears to be separate from others. That appearance of separation is only possible with the self-identification of a presence of consciousness with the central form of a body. That appearance of separation, self and other is only an illusion, which comes to an end when that world of form disappears. Ultimately, nothing exists, there is no separation, no self, and no other. The void is 'all-one' and alone.

Mu-mon describes the path of return with the gateless gate paradox, which expresses that the divided consciousness of an observer can only be present for a world of form at a point of view, or return to its true undivided, formless state:

The great path has no gates,
Thousands of roads enter it.
When one passes through this gateless gate,
One walks the universe alone.

It is tempting to attribute intentionally to the void, but that is just not the nature of the void. There is no intentionality in the void, only potentiality. The void creates a world because it can. That is its potentiality. Everything that can possibly appear to happen in the world actually does appear to happen in some world. As expressed in *The Once and Future King*: 'Everything not forbidden is compulsory'. All actions of the void are impersonal, as is the true nature of consciousness. Intentionality is only about actuality.

Intentionality only arises with the flow of energy through the world, the organization of information into the form of a body, and the emotional actions of a body.

All things arise from Tao
They are nourished by the energy of Tao
They are formed from matter
They are shaped by environment

Tao in the world is like a river flowing home to the sea

Tao follows what is natural

A body is an animated form of information that naturally arises in the world. That form is coherently self-replicated in form over events in the flow of energy, as the behaviors of that form are enacted. Only the coherent flow of energy through a body allows for the coherent self-replication of the form of that body. We perceive those emotional actions as body feelings. It is not possible to understand the self-identification of a presence of consciousness with the form of a body without a discussion of emotional expressions.

There are two kinds of emotional expressions that are inherently related. The first kind of expression is an emotional attachment, and the second kind is a self-defensive expression. They are related since self-defensive expressions only arise with emotional attachments. It only makes sense to defend attachments. Without an attachment to something, there is nothing to defend. Attachments come first, followed by self-defensive expressions.

All complex organisms must attach themselves to the parent organism early in their development. The reason for this is simple. A developing organism feeds upon the parent organism. The coherent organization of a biological organism only develops through a process of eating. Coherent organization only develops through a process of biological symmetry breaking, which is dependent on a process of eating, as high potential energy molecules are added to the body, burned within the body, and disordered kinetic energy is radiated away. All body development and growth requires a process of eating.

This is as much the case for plants as for animals. The only difference is animals eat other biological bodies, and plants eat photons. The process of photosynthesis inside a plant only goes forward as a high energy photon from the sun is absorbed, and lower energy photons are radiated away into the environment. The lower energy photons are radiated away as heat into the cool night sky. This process can only go forward as heat flows from a hotter to a colder body. That flow of energy is associated with an increase in entropy. An increase in disordered information arises since more infrared photons are radiated away into the cooler sky each night than the number of yellow photons that arrive each day from the hot sun. The reason more infrared photons are radiated away is the energy of each photon is quantized as $E=h\nu$, and energy is conserved. The same amount of energy that arrives is radiated away. This process is only possible due to gravitational clumping of matter and energy that occurs after the big bang event, and allows for a hot sun to form in the sky by day, and a cool dark sky by night (Penrose 2005, figure 27.9).

The same kind of thermodynamic processes occur as an animal eats another biological body. High potential energy molecules are added to the body of the animal, burned within the body, and heat is radiated away. These thermodynamic processes allow for development of coherent organization through symmetry breaking. All development, growth and survival of biological organisms requires these thermodynamic processes. A biological body only develops, grows, and survives if it eats other bodies.

The development of a biological body depends on development of coherent organization, which only occurs through a process of biological symmetry breaking, and is dependent on a process of eating. Bodies must eat each other in order to develop and survive. Body survival is inherently the self-replication of the form of that body over a sequence of events, while the behaviors of that body are enacted. Self-replication of form is the nature of how information becomes coherently organized in a distinct thermodynamic phase of organization, which gives rise to the distinct macroscopic appearance of that form. Only coherent organization allows for self-replication of form over a sequence of events. These thermodynamic processes in biological bodies are dependent on a process of eating.

A biological body only develops and survives if it eats other bodies. Early in its development, a developing organism only develops if it feeds upon the parent organism. The process of eating is a natural aspect of the life of a body. Emotional expressions naturally arise with the life of a body. Every expression of desire is a kind of hunger, and expresses the desire to eat something. Body survival is not possible without these kinds of expressions. The expression of the desire to eat is fundamentally about the desire to live. Coherent self-replication of the form of a body is only possible with the coherent flow of emotional energy through that body, which allows for the self-replication of that form over a sequence of events, and for body survival (Damasio 1999, 39, 138).

These kinds of emotional expressions are at the heart of emotional attachment. Every expression of desire is a kind of hunger that expresses the desire to eat something. That desire to eat is expressed from the very beginning of life of the developing organism, as it attaches itself to the parent organism, and feeds upon the parent organism. That desire to eat is expressed by the newly born organism, but is also expressed by the fetal organism, and even by the unfertilized egg cell. Every biological organism, whether multi-celled or single-celled, expresses the desire to eat, and to live. It is not even possible to define a living biological organism without the expression of that desire (Damasio 1999, 51, 136).

Emotional attachments only arise as desires are expressed and satisfied. The satisfaction of a desire feels good, which is described as pleasurable. The most basic desire that is expressed is the desire to eat, which feels good as that desire is satisfied. The frustration of a desire feels bad, which is described as painful (Damasio 1999, 77).

Emotional attachments only arise as desires continue to be expressed and satisfied over a sequence of events. We might say that an attachment to something only arises from the perpetuation of that expression of desire and its satisfaction. Attachments arise as desires are expressed in repetitive cycles of the expression of desire, and those desires are satisfied. The satisfaction of a desire feels good, which perpetuates the expression of those repetitive cycles. The most basic example of attachment is the desire to eat something. In this sense, the attachment of a developing organism to its parent organism naturally arises as it feeds upon the parent organism.

How can we understand the expression of desire, and the satisfaction and frustration of desires, in energetic terms? The expression of desire is inherent to the way the organism is coherently organized. It is not even possible to discuss a distinct living organism unless the form of that organism is coherently self-replicated over a sequence of events, which requires the coherent flow of emotional energy through that organism.

That coherent flow of emotional energy through the organism is the expression of desire, which allows for the self-replication of the form of the organism while the behaviors of the organism are enacted over a sequence of events. The most basic desire that can be expressed in the desire to eat something. Without that emotional expression, the coherent organization of the organism cannot develop, and is not perpetuated over those events.

If the desire to eat something is satisfied, the coherent flow of emotional energy through the body of the organism comes into alignment with the flow of energy through that other thing. That is the nature of a process of eating, as that thing is absorbed within the body of the organism. A process of eating by an organism only arises with the alignment of the flow of energy between the organism and the thing that is eaten. As that flow of energy comes into alignment, feelings of connection are expressed by the body of the organism. Those feelings of connection feel good, which is perceived as pleasure. If the desire to eat something is frustrated, the coherent flow of emotional energy through the body of the organism goes out of alignment with the flow of energy through that other thing. That other thing is not absorbed within the body of the organism. As that flow of energy goes out of alignment, feelings of disconnection are expressed by the body of the organism. Those feelings of disconnection feel bad, which is perceived as painful.

Emotional attachments only arise as desires are expressed and satisfied, and feelings of connection are expressed, which feel good. The attachment is perpetuated as those desires are subsequently expressed and satisfied in future events. The attachment arises as the flow of emotional energy through the body of an organism comes into alignment with the flow of energy through some other thing, such as a developing organism that attaches itself to the parent organism as it feeds upon the parent. Inherent in that attachment is a process of eating, which allows for the coherent self-replication of the form of that body. The attachment can only arise as that desire to eat something is expressed and satisfied.

If the desire to eat something is frustrated, feelings of disconnection are expressed, which feel bad. Self-defensive expressions, like fear and anger, arise out of that frustration. Self-defensive expressions inherently defend the survival of the body. Body survival is dependent on the process of eating and the satisfaction of that desire. Self-defensive expressions naturally arise if those desires are frustrated, and body survival is threatened. Self-defensive expressions are a natural aspect of embodied life in an inherently dangerous world, where bodies must eat each other in order to survive.

Self-defensive expressions inherently defend the survival of a body. These kinds of self-defensive expressions are a necessary aspect of embodied life in an inherently dangerous world where bodies must eat each other in order to survive. With the expression of desire, the body moves toward those things that promote its survival. With the expression of fear, the body moves away from those things that threaten its survival. With the expression of anger, the body moves against those things that threaten its survival. Self-defensive expressions inherently defend the attachments that are formed as desires are satisfied. The survival of the body depends upon perpetuating those attachments.

These self-defensive emotional expressions are inherent in how the body is coherently organized. The body is coherently self-replicated in form as behaviors are enacted over a sequence of events. The coherent flow of emotional energy through the body allows for that self-replication of form. Only a coherently organized body can express self-defensive emotions, which inherently defend the survival of the form of that body, and inherently are limited to that form. The flow of energy through a body can come into alignment with the flow of energy through other things, and create an emotional attachment to those forms, but if the body is to survive and self-replicate its form over a sequence of event, that coherent flow of energy at most can become limited to the body itself.

As desires are frustrated, the flow of energy through the body goes out of alignment with the flow of energy through some other thing, but if the body is to survive, that flow of emotional energy must remain coherently organized as it flows through that body. Self-defensive expressions inherently arise from that coherent flow of emotional energy through the body itself. Body feelings of connection arise as desires are satisfied, and body feelings of disconnection arise as desires are frustrated. Self-defensive expressions arise with the self-limitation of that flow of emotional energy within the body itself.

So far, emotional expressions have only been discussed in the sense of the emotional flow of energy through the body, and how that flow of energy can come into or go out of alignment with the flow of energy through other things. If there is alignment, feelings of connection are expressed. If there is no alignment, feelings of disconnection are expressed. Attachments naturally arise with the satisfaction of desires. Self-defensive expressions naturally arise with the frustration of desires, and defend attachments. We might say that the ultimate attachment is to the body itself, since the ultimate level that the flow of emotional energy can become self-limited to is to a coherently organized body. In this sense, the ultimate defense of an attachment is the defense of body survival.

This discussion of the energetic nature of emotional expressions is as far as we can go without introducing the concept of consciousness. Feelings of connection feel good, and feelings of disconnection feel bad, but who perceives those body feelings? Those body feelings inherently represent the way emotional energy flows through the body. Body feelings are somehow represented in the mind, but who is the perceiver of the mind?

The discussion that follows is really no different than the discussion that can be found in the mainstream neuroscience literature, except in one important aspect. In large part, this discussion is the same argument Damasio makes about the nature of the self-concept in *The Feeling of What Happens*. Damasio's argument is really not new, and is largely the same argument found in the psychoanalytic literature, usually referred to as Object Relations Theory. What Damasio has done is to give these ideas a sounder scientific basis. Damasio describes how body feelings are represented in the mind, and how the self-concept arises from emotional relationships between a body-based self-image and the image of another (Damasio 1999, 133). What Damasio has not done is explain the nature of the observer of the mind. The one important aspect of the discussion that Damasio left out, and that all neuroscience discussions leave out, is about the nature of the observer.

The natural way to understand the observer of the mind is with the holographic principle. The mind displays an entire world perceived by the observer. The mind arises on a viewing screen, which is an event horizon that encodes quantized bits of information. The event horizon arises from the central point of view of the observer. The event horizon arises in empty space, and the observer arises at a focal point of perception. All the information for the observable world is encoded on the viewing screen. That information includes the central form of a body. Organs of sensory perception in that body relay information about that entire world, which includes external sensory perceptions of that world, and internal perceptions of emotional body feelings. All of those perceptions are like images that are projected from the viewing screen to the central point of view, where the consciousness of the observer is present. That observer only feels like it is embodied within that body, as it perceives the emotional body feelings expressed by that body.

This state of affairs cannot be stressed strongly enough. Self-identification of a presence of consciousness with the form of a body is only a feeling. A feeling is only a perception. That observer only identifies itself with the form of a body since it really feels like it is embodied in the form of that body as it perceives the emotional body feelings expressed by that body. That body-based self-identification is inherently emotional in nature.

How is self-identification even possible? We might say that perception is recognition. A presence of consciousness recognizes itself in the forms of information that it perceives since those forms are emotionally animated over a sequence of events in the flow of energy. The perception of emotional actions leads to emotional self-identification with the form of a body, which is another way to say the observer really feels like it is embodied in the form of a body that expresses those emotional actions. Those emotional actions are perceived as body feelings. The observer recognizes itself in those actions.

How can a presence of consciousness recognize the forms of information it perceives? Forms only arise through a process of symmetry breaking, as bits of information tend to align with each other and form bound states. The symmetry broken is the symmetry of empty space. In a physical sense, the true nature of consciousness is empty space. A form is only self-replicated in form over a sequence of events due to the coherent flow of energy through that form. The coherent flow of energy through the form of a body is perceived as emotional expression, which allows for self-replication of form while behaviors are enacted. Emotional actions allow for recognition, which is another way to say 'to act is to give meaning'. Emotional meaning is given to emotional actions. That is another way to say 'feeling is believing', just as 'seeing is believing'. All perceiving is believing. Beliefs are mentally constructed concepts that only arise with emotional actions. Mental concepts are only believable due to the meaning given to them as they are perceived. The meaning given to an action only arises within an emotional context. That emotional context only arises from the way things are connected to each other, due to coherent organization and alignment of information. The meaning given to an action arises with the alignment of information, which arises through a process of symmetry breaking. The symmetry that is broken is the symmetry of empty space. Consciousness recognizes itself in all of its actions, since all of its actions arise from its true nature.

This is the best scientific explanation we will ever have for the nature of consciousness. We are trying to explain a mystery that is transcendent of all explanations. No scientific concept can ever explain the nature of consciousness, since consciousness itself is what perceives and understands those concepts, and it can never be reduced to any concept it perceives. This is what the incompleteness theorems prove. All concepts can be reduced to the way information is encoded in the world, and the way that information becomes coherently organized into form, as energy flows through the world. The consciousness that perceives and recognizes those coherently organized forms of information can never be reduced to a form of information it perceives. A presence of consciousness is always outside that world of form, as it perceives and recognizes those forms of information.

The mystery of consciousness is how it effects actions, which is the mystery of free will. In a state of non-interference with the normal flow of all things, there is no expression of personal will, free or not, since only the path of least action is taken. The quantum state of the world constantly branches, and every event is a decision point where the path branches into alternative paths. The path of least action is only the most likely path in the sense of quantum probability. In the sense of many worlds, every path is taken, and no decision is ever made. In the sense of non-interference with the normal flow of all things, only the path of least action is taken, and no decision is ever made. The mystery is how consciousness effects the decisions that are made. The only possible explanation is summarized with 'to act is to given meaning'. Actions arise from the way energy flows. Every event is a decision point where the quantum state branches into alternative paths. An action is what connects two events on the path. Every action that is observed is given meaning to by the presence of consciousness that observes that action. The nature of action and meaning are intrinsically related.

By connecting two events on the path, action creates the context within which meaning is given, which is to say 'to act is to give meaning'. Meaning arises from the way things are connected to each other, which only arises with actions. The mystery of free will is related to this connection. The meaning given to any action has an effect on future events, since action connects two events on the path, and the path that is followed in some way arises from the meaning that is given to the actions. Every event is a decision point where the path branches into alternative paths, and the meaning given to any action that connects two event on that path effects which of those alternative paths will be chosen at that decision point. But the expression of will is only free to the degree one is free to give meaning to the expression of actions. Since 'perceiving is believing', there is really no freedom at all. The only real freedom one has is to disbelieve false beliefs. The meaning one gives to beliefs that arise from actions is to believe them. Disbelief is the meaning that one gives to an action when one sees the inherent falseness of a belief that arises from that action. Since disbelief leads to a state of non-interference with the normal flow of all things, disbelief naturally leads to a state of non-doing and non-action, which in some sense is the willingness to choose not to choose.

We have just introduced the concept of belief. A belief is a mentally constructed concept that only arises with emotional action. Only the presence of consciousness for that mind perceives and recognizes that belief, and gives emotional meaning to that belief. That presence of consciousness is the believer. The perceiver of the mind is the believer of all the beliefs emotionally constructed in that mind. There is something inherently circular about this explanation. The perceiver recognizes itself in that perception as it believes that belief. The nature of belief is reflected in every self-concept. The perceiver only identifies itself with a body-based self-concept since it really feels like it is embodied as it perceives the emotional body feelings expressed by that body, and feeling is believing. In this sense, self-identification with the form of a body is a belief. The perceiver believes it is embodied. The circular nature of belief reflects the holographic principle, since the mind is only a viewing screen that arises in empty space, as an observer arises at a point of view. Both that observable world and the observer of that world arise from the void.

The discussion of belief that follows is no different than the discussion Damasio gives in *The Feeling of What Happens*, except in one important aspect. Damasio does not identify the nature of the observer of the mind, even as Damasio describes how that observer becomes emotionally self-identified with a body-based self-concept that is emotionally constructed in the mind. The holographic principle naturally identifies the observer with a presence of consciousness that arises at a point of view in empty space as the mind arises on a viewing screen. It is not possible to say what that observer is, except to say the observer arises from the void, as the world the mind displays arises from the void.

Damasio points out there is emotional expression in every expression of belief, and in every concept. Beliefs only arise with emotional relationships between self and other, and implicitly require our belief in a body-based self-concept (Damasio 1999, 21, 133). Every concept of self and other constructed in the mind relates a body-based self-concept to the concept of some other thing with an emotional body feeling. We perceive body feelings as we perceive the emotional flow of energy through the body inherent in the enactment of behaviors. Only emotional expressions allow for self-replication of form, and maintain the organization of the body. (Damasio 1999, 39, 138). Those emotional expressions are inherently self-defensive in nature, as body survival is defended. A body only develops and survives through a process of emotional attachment. Attachments are defended if threatened, just as body survival is defended if threatened.

A self-concept is only emotionally constructed and self-replicated in a mind with these self-defensive expressions, as the concept of self is emotionally related to the concept of other things that appear in the world. The mental construction of a self-concept only arises as an emotional projection to past or future events, as a body-based self-image is emotionally held in mental imagination, and is related to the images of other things held in mental imagination (Damasio 1999, 133). Holding of images in mental imagination is the nature of what we call memory and anticipation of events, and is inherently emotional in nature. The holding of a body-based self-image in mental imagination is inherently self-defensive in nature. The construction of any mental concept is emotional in nature, as an image is emotionally held in mental imagination and related to other images.

Memory only arises with the development of coherent organization of information. Memory of events only develops through a process of conditional probability that arises with the observation of events. In terms of a quantum state of potentiality, every event is a decision point where the path branches. In the sense of a quantum state reduction, memory only arises through a decision process that truncates the probability distribution and alters conditional probabilities. This decision process occurs with every observation of events. In this sense, memory of events and the development of coherent organization only occurs with observation of events. In the many world sense, the potentiality for all possible events co-exist as a superposition of all possible paths, and all alternative paths co-exist in an ensemble of all possible worlds. An observer is present for every possible event, but that observer only has access to those memories that arise on a particular path, since all the alternative paths of all the parallel worlds do not interfere with each other. In either case, only observation of events allows for a decision process that leads to the development of coherent organization, which underlies all memory of events.

There is another aspect of coherent organization that needs discussion. A coherently organized system only arises through symmetry breaking. The coherently organized form of that system takes on a distinct macroscopic appearance that only arises in a coherently organized thermodynamic phase of organization. Development of coherent organization is inherently thermodynamic in nature. Self-replication of the form of that system only arises within that coherently organized phase of organization. Self-replication of form is nothing more than maintaining that macroscopic appearance while the microscopic details change. Inherent in the development of such a coherently organized phase of organization is the concept of a thermodynamically meta-stable state (Davies 1977, 174).

Such meta-stable states are common in nature, like super-cooled liquid water that can rapidly transition into the form of ice. As ice forms, heat is radiated away from the system. The system of water molecules in the form of ice is more ordered, since the molecules take on a more orderly arrangement of position in space as water freezes. The amount of disordered information, or entropy, is less for the system of water molecules in the form of ice than for the form of liquid water, even at the same temperature. The entropy of the system decreases as water freezes. How is this possible? As water freezes, heat is radiated away into the environment, and heat is disordered kinetic energy. The total entropy of the system and the environment increases due to the heat radiated away.

A key aspect of such thermodynamically meta-stable states is the possibility of amplification of a microscopic signal. Such a microscopic amplification occurs when a photon interacts with a retinal cell, and a photo-chemical reaction occurs inside the retinal cell. That photo-chemical reaction is no different in kind than the photo-chemical reactions that occur on a photographic plate. The atoms in that photographic plate are in a thermodynamically meta-stable state, and as they interact with photons they transition to a more stable state, just like super-cooled water that freezes into ice. The image that arises on the photographic plate as a consequence of those photo-chemical reactions literally 'freezes out'. As the image freezes out, there is a local decrease in entropy, and the encoding of information on the photographic plate. The only reason this is possible is due to the heat that is radiated away from the photographic plate into the environment as the photo-chemical reactions occur. But before that information can become encoded on the photographic plate, the atoms in the plate must become organized into a meta-stable state, which has the potentiality to thermodynamically encode information. The encoding of information with the distinct macroscopic appearance of an image only arises through the irreversible process of the amplification of a microscopic signal. That process only appears irreversible due to the heat radiated away into the environment. That irreversible process only goes forward with construction of a thermodynamically meta-stable state.

That irreversible process only arises as the quantum state of potentiality branches into alternative paths. Every event is a decision point where the path branches. In either the sense of a quantum state reduction or many worlds, that irreversible process only arises with the observation of an event. But the many world interpretation is the more natural thermodynamic interpretation, since all the alternative paths co-exist in an ensemble of all possible worlds. Irreversibility is only a characteristic of those worlds in which the total entropy of the world appears to increase over the course of time (Davies 1977, 172). Irreversibility arises with an ordered initial state of such a world (Feynman 1963, I 46-7). A key aspect of the encoding of information in such a world is the construction of thermodynamically meta-stable states that allow for the irreversible amplification of microscopic signals. This is the natural way to understand the nature of memory and the development of coherent organization in any system (Davies 1977, 174). Memory only arises from the order inherent in the initial state of that world (Feynman 1963, I 46-7).

A meta-stable state is characterized by a potential barrier. A carbohydrate molecule is a meta-stable state of high potential energy that transitions to a more stable state of lower energy as it burns and heat is radiated away. The molecule is in a meta-stable state due to a potential barrier, which prevents the spontaneous burning of the molecule. The electromagnetic repulsion between protons is another example of a potential barrier that prevents spontaneous fusion. A potential barrier is like a hill between two valleys that must be climbed before the transition to the more stable state can occur. The meta-stable state is held in a higher potential valley, and the more stable state is a lower valley that is only reached if the hill between those two valleys is climbed. That hill is only climbed if kinetic energy is added to the molecule, which is why heat must be applied before the molecule will burn. Burning to a more stable state occurs as heat is radiated away, and the system settles into the lower potential valley. If that heat is not radiated away, the system has too much kinetic energy and the potential barrier can be climbed again, and the system can return to the higher potential valley and that meta-stable state.

A meta-stable state also explains why the transition to the more stable state occurs with a cascade effect, or like an avalanche. Burning begins slowly, but accelerates to a faster and faster rate, just like water flowing down a river. An avalanche occurs suddenly, when the downward force of gravity overcomes the cohesive force of a snow pack that holds it together and to the mountainside. Once the fall begins, it accelerates.

There is another way the transition can occur, which is a tunneling event. Due to quantum uncertainty in position and momentum, the system can tunnel through the potential barrier and reach the more stable state. But if the potential barrier is very large, the decay time for that tunneling event to occur is very long, and is unlikely to occur. The more likely way for the transition to the more stable state to occur is to add kinetic energy to the system, which allows the system to climb the potential barrier. In either case, as the system transitions to the more stable state, heat is radiated away.

There is even a third way the transition can occur, which is the effect of a catalyst. As we add a catalyst to the system, we lower the potential barrier, and the transition to the more stable state is more likely to occur, either because the system has enough kinetic energy to climb the lowered potential barrier, or due to a tunneling event that is more likely to occur with a lowered potential barrier. In either case, heat is radiated away as the system transitions to the more stable state, and the system appears to burn.

The addition of a catalyst to the system is the addition of potential energy. The potential barrier is lowered since we are adding negative potential energy. The addition of negative potential energy to a system like a carbohydrate molecule with a potential barrier allows that system to burn. But how was that potential barrier created in the first place? The potential barrier was constructed due to the addition of positive potential energy. The construction of the potential barrier in a carbohydrate molecule was constructed through the process of photo-synthesis, which is like the addition of positive potential energy to the system. That process only goes forward as the directed kinetic energy of a photon is converted into positive potential energy. But even the process of photo-synthesis requires that disordered kinetic energy is radiated away from the system into the environment. The process of constructing the carbohydrate molecule leads to an overall increase in entropy in the world just as much as the process of burning the carbohydrate molecule.

We can construct a carbohydrate molecule through the process of photosynthesis, as the directed kinetic energy of a photon is converted into positive potential energy. We can burn that carbohydrate molecule with the addition of the negative potential energy of a catalyst. In either case, these processes can only go forward as heat is radiated away into the environment, and the total entropy of the world increases. All we are really doing is constructing a potential barrier, and then deconstructing that potential barrier. That potential barrier is the nature of a meta-stable state, which is like a hill that must be climbed before the transition to the more stable state can occur. The climbing of a hill in order to reach a more stable state is metaphorically represented in the myth of Sisyphus. The only way that more stable state is reached is if the system burns, and heat is radiated away. If there is no burning, the system is doomed to remain in the meta-stable state.

The nature of biological symmetry breaking only arises with the addition of potential energy to a system, which is a process of eating. The addition of positive potential energy constructs the potential barrier of a meta-stable state, which is only deconstructed through a process of burning. The confusing aspect of the process of burning is that it can go forward with the addition of negative potential energy, which deconstructs the potential barrier. The construction of a potential barrier in a meta-stable state is a key aspect of the encoding of information in any system. Only thermodynamically meta-stable states allow for the irreversible amplification of microscopic signals. Only this process allows for the development of coherent organization of information in any system. This is the natural way to understand how information and memory are encoded in a mind.

A key aspect of the development of coherent organization and memory in a mind is the development of a self-concept. In some sense, the self-concept is mentally constructed as a meta-stable state of high potential energy, which is characterized by a potential barrier. In some sense, that potential barrier is the expression of self-defensiveness. These emotional expressions inherently defend body survival, and allow for development of a body-based self-image that is emotionally held in mental imagination, which Damasio calls a proto-self (Damasio 1999, 154). This proto-self is a body-based self-image that is continuously constructed in all states of self-consciousness. That construction process arises with the maintenance of body stability (Damasio 1999, 141). In a thermodynamic sense, that body-based self-image is self-replicated in the same coherently organized phase of organization, and the expression of self-defensive emotions maintains that coherent organization. That self-image is only a macroscopic appearance that arises from the way information is coherently organized at a microscopic level.

Simply stated, it is not possible to have a self-concept without the self-replication of such a body-based self-image. A self-concept only arises as that body-based self-image is emotionally related to the images of other things held in memory (Damasio 1999, 169). The nature of those emotional relationships are body feelings, which represent the flow of emotional energy through the body. This process occurs on a moment-by-moment basis, and only depends on short-term memory of events (Damasio 1999, 112). Damasio calls this process core self-consciousness. If there are also long-term, or autobiographical memories, then there is also a sense of an autobiographical self (Damasio 1999, 174).

Memory arises with the development of coherent organization in a developing organism. The possibility of constructing a self-concept requires a tremendous degree of coherent organization of information (Damasio 1999, 175). To fully develop a self-concept, a body-based self-image must be emotionally held in mental imagination over a sequence of events, and must be emotionally related to other images held in mental imagination. The holding of those images in mental imagination requires the self-replication of the form of those images over a sequence of events, which can only arise with development of coherent organization in the developing organism. But the memory of events is not the same as the consciousness that perceives those events. This is the mistake that Damasio makes. Consciousness can be present for every event even if there is no memory of prior events. Consciousness can be present for every event even if there is no self-concept.

The mistake Damasio makes is that he assumes the physical world is the only reality. Within that physical world, there is no place for consciousness, and no place for the observer. He explicitly states: "There is no external spectator". He describes "the images that constitute the narrative" and "are incorporated in the stream of thoughts", and states "images in the consciousness narrative flow like shadows". He uses "the metaphor of the movie-in-the-brain" and states those images "are *within* the movie" (Damasio 1999, 171). It is as though the images in the movie perceive themselves, and that a two dimensional image animated on the screen is its own observer, which is a paradox of self-reference.

In one sense Damasio is absolute right. There is no place for the observer within physical reality. There is no physical explanation of consciousness, as he correctly points out. But the idea Damasio has of reality is too limited. He assumes physical reality is the only reality. In this sense, he is stuck in the nineteenth century, with his outdated ideas of absolute space and time and classical determinism. He assumes that reality only consists of matter and energy within some pre-existing space and time. The holographic principle, as embodied by all modern unified theories, explicitly demonstrates these assumptions are incorrect. Physical reality is like a movie of images that are displayed on a screen and observed by an observer, but that screen is only an event horizon that arises within empty space. That screen encodes information for that physical world. That screen always arises from the central point of view of the observer, which is 'external' to the screen.

This description of the observer is a metaphysical description, not a physical description. The observer is only describable in physical terms as a point of view in empty space. It is impossible to take the meta out of physics since it is impossible to take the observer out of physics. Simply stated, without the observer there would be no physical world.

The development of a self-concept in a child demonstrates the necessity of organism development prior to the development of a self-concept. A self-concept only appears in a young child around two years of age, with the development of language. Prior to this age, the child exhibits no evidence of a self-concept (Damasio 1999, 175). Both the development of a self-concept and the development of language capability depend on memory, which requires the development of coherent organization of information in the developing organism. Consciousness can be present for that developing organism prior to the development of memory, prior to the development of language capability, and prior to the development of a self-concept. Prior to the development of memory, that presence of consciousness has no memory of prior events. Prior to the development of a self-concept, that presence of consciousness has no sense of self.

The development of a self-concept in a young child only arises through a process of emotional attachment. The emotional attachment of the child to the parent is necessary for the child to develop, for the simple reason the parent must feed the child for that development to occur. All development, growth and survival is dependent on a process of eating. Only the addition of potential energy to the organism allows for the development of coherent organization through a process of biological symmetry breaking.

A self-concept only develops within the emotional context of a society. A human society is only a collection of emotional relationships between people. A human society only holds together due to emotional attachments between different people in that society. Emotional attachments are the nature of attractive interactions between different people that hold a society together. The fundamental nature of those attachments are the emotional relationships between self and other. Attachments between people only form as desires are satisfied. Attachments are only defended as desires are frustrated. In energetic terms, an attachment forms as the flow of energy through one body comes into alignment with the flow of energy through another body. Attachments are self-reinforcing in nature since those feelings of connection feel good. Attachments are defended since feelings of disconnection feel bad. Attachments begin with the birth of a child into a family, and persist throughout the life of that person as long as that person remains a part of a society. Without those emotional attachments, a society cannot hold together.

All complex organisms must attach themselves to the parent organism early in their development, but to fully mature they must detach themselves. That development is arrested in an immature state if there is a failure of detachment. We recognize that immature state as a state of dependency on others. A person only remains a part of a family or a society if there is a failure of detachment. A person only breaks free of their bondage to a family or a society through a process of detachment, which is the only process that leads to autonomy and self-reliance.

All the great spiritual writings of the world describe the process of spiritual awakening as a process of detachment. This cannot be stressed strongly enough. Spiritual awakening is only possible through a process of detachment. The fundamental reason for this state of affairs is that a self-concept only arises through a process of emotional attachment.

A self-concept only develops in a young child through a process of emotional attachment to the parent. Without that emotional attachment, a self-concept cannot develop. A self-concept only develops with the development of memory and coherent organization in the developing child, and is typically expressed through language. A self-concept is mentally constructed as a body-based self-image is emotionally held in mental imagination, and is emotionally related with body feelings to other images held in mental imagination. The holding of images in mental imagination is the nature of memory, which only arises with the self-replication of the form of those images over a sequence of events. The nature of the emotional expressions that relate those images only arise with emotional attachments that arise as desires are expressed and satisfied, and with self-defensive expressions that arise as desires are frustrated and attachments are defended. Without emotional attachments, the expression of a self-concept is not even possible.

This is the fundamental reason the process of spiritual awakening requires a process of detachment. By its very nature, spiritual awakening is the end of the mental construction and emotional expression of a self-concept. Spiritual awakening is the awakening of a presence of consciousness to its true nature. That presence of consciousness is the observer of the mind. Without the emotional expression of a self-concept in its mind, the observer of the mind no longer emotionally identifies itself with a body-based self-image.

Self-identification is only possible since the observer really feels like it is embodied in that body as it perceives the emotional body feelings expressed by that body. Those body feelings are inherent in every self-concept that emotionally relates a body-based self-image to the image of some other thing. This is the only process that allows for the self-identification of the observer with the form of a body-based self-image. The process of spiritual awakening brings to an end that emotional self-identification with a self-image.

All the great spiritual writings of the world tell us the process of spiritual awakening only goes forward with a process of detachment, which finally brings to an end the emotional expression of a self-concept. That detachment process is the only way the observer of a mind can detach itself from its self-concept. This is explicitly stated in the Tao:

He who is attached to things will suffer much

The sage stays behind, thus he is ahead
He is detached, thus at one with all
Through selfless action, he attains fulfillment

The satisfaction of desires feels good, which creates an emotional attachment of the observer to something in the world, and perpetuates the expression of that desire. As desires are satisfied, the flow of emotional energy through the body comes into alignment with the flow of energy through some other thing in the world, and feelings of connection are expressed. The frustration of desires feels bad. As desires are frustrated, the flow of emotional energy through the body goes out of alignment with the flow of energy through some other thing, and feelings of disconnection are expressed. Out of that frustration, the desire to possess things, control things, and force things to satisfy desires is expressed.

He who grasps loses

Nothing ever wants to be possessed or controlled, and eventually all things resist those emotional attempts at control. The desire to hold onto, possess, and control things will ultimately turn all things into pain-giving things that frustrate desires, since nothing wants to be possessed or controlled. An observer that clearly sees the futility of its desire to control things is ultimately willing to let go of those things, and receive nothing in return, since it would rather be pain free than continue to hold onto a pain-giving thing.

All can know good as good only because there is evil
For having and not having arise together

Misfortune comes from having a body
Without a body, how could there be misfortune?
Surrender yourself humbly; then you can be trusted to care for all things
Love the world as your own self; then you can truly care for all things

The willingness to let go of attachments leads to autonomy, which is the only process that allows for the development of self-reliance. Attachments perpetuate a state of immaturity and dependency on others. Development of autonomy is the process of letting go and growing up. The willingness to let go of attachments always feels like something is dying inside. What dies? Ultimately, the illusion of a self-concept dies. The illusion of being a person in the world dies. Even that illusion does not really die as long as the body lives. Only belief in that illusion dies. False belief in self and other dies. A self-concept is a mentally constructed belief that arises in emotional relationship with the concept of other.

That belief dies when it is no longer believable. Belief comes to an end in the desireless state. Without belief, the self-concept is only a character role that we play, like an actor on a stage. The irony is that to know the truth, all desires must die, including the desire to know the truth. The self-concept only dies through a self-destructive process, which only begins if the self-concept is examined, and its falseness is exposed and clearly seen. That examination turns a self-concept into a pain-giving thing, which is the reason the observer is willing to let go of its attachment to it, and receive nothing in return. This self-destructive process only goes forward with willingness to suffer ego death rather than live the life of a lie.

It is more important
To see the simplicity
To realize one's true nature
To cast off selfishness
And temper desire

If nothing is done, then all will be well

Empty yourself of everything

Without form there is no desire
Without desire there is tranquility

Therefore the sage seeks freedom from desire

The ultimate desire is the desire to live, and the ultimate fear is the fear of death. Only a body can express desires. Only those emotional expressions allow for the self-replication of the form of the body while behaviors are enacted. A body is a coherently organized bound state of information that self-replicates its form over events in the flow of energy. A body only develops with the development of that coherent organization, and body death occurs with the loss of that organization. The nature of the flow of energy through the world is that all forms eventually become disorganized, since all energy flows in the sense of thermodynamics. That universal flow of energy allows for transformation of form into new form. Emotional expressions are inherently self-defensive in nature as they defend the survival of the body, and maintain that organization over a sequence of events.

A self-concept is only emotionally constructed and self-replicated in a mind with those self-defensive expressions, as the concept of self is emotionally related to the concept of other things that appear in the world. The mental construction of a self-concept only arises as an emotional projection to past or future events, as a body-based self-image is emotionally held in mental imagination, and is related to the images of other things that appear in the world (Damasio 1999, 133). The holding of images in mental imagination is the nature of memory and anticipation of events, and is inherently emotional. The holding of a body-based self-image in mental imagination is inherently self-defensive in nature.

Self-defensive expressions naturally arise with emotional attachments. It only makes sense to defend those attachments. All complex organisms must attach themselves to the parent organism early in their development, but to fully mature they must detach themselves. That development is arrested in an immature state if there is a failure of detachment. That failure of detachment only arises with an exaggeration and distortion of the normal expression of self-defensiveness in mental imagination.

The failure of detachment only arises with an ego, or the mentally constructed concept of self and other, which perpetuates an immature state of dependency. Exaggerated and distorted self-defensive expressions only arise in mental imagination, as a body-based self-image is emotionally held in mental imagination through emotional projection to past and future events, and is emotionally related to the images of other things in the world. Those emotional relationships are the nature of self-referential thoughts, which only arise with the memory and anticipation of events, as images are held in mental imagination. The body no longer just responds to threats to its survival in the moment, but also responds to imagined threats to its survival as constructed in mental imagination. Each self-referential thought is a stimulus for another self-defensive emotional response in the body, which leads to the construction of more self-referential thoughts. Self-defensive expressions become exaggerated and distorted in mental imagination, and create a vicious cycle, just like the distortion and amplification that arises in an out-of-control positive feedback loop. Self-defensive expressions become amplified and distorted in mental imagination due to the self-reinforcing positive feedback nature of mental imagination.

Self-defensive expressions inherently defend the survival of the body. Self-defensive expressions are a necessary aspect of embodied life in an inherently dangerous world where bodies must eat each other in order to survive. With the expression of fear, the body moves away from those things that threaten its survival; with the expression of desire, the body moves toward those things that promote its survival; and with the expression of anger, the body moves against those things that threaten its survival. Emotional conflicts naturally arise in mental imagination as something is initially perceived as desirable, but then turns into something that is perceived as threatening.

The expression of power is also an aspect of self-defensive expression. The expression of power is the nature of grandiosity, just as vanity is the expression of self-love, and narcissism is the expression of being in love with one's own self-image. People do things because they like the expression of power. Expression of power makes one feel powerful.

The sense of entitlement to special treatment arises with this sense of self-importance. The accumulation of wealth is an aspect of the expression of power, as is the desire to force one's will upon others, and force others to satisfy one's desires. The expression of power is the expenditure of energy applied over time, just as work is the application of a force applied through a distance, and is all about the flow of energy through the world. The expression of power is about the desire to force others to satisfy one's own desires, which is just as self-defensive as the expression of anger, which is the desire to attack and destroy others, and the expression of fear, which is the desire to run away from others. These expressions only defend the survival of a body from the threats of other bodies in an inherently dangerous world where bodies must eat each other in order to survive.

A body is born as information is coherently organized into the form of a body. A body lives as long as that form is coherently self-replicated in form. A body dies with the loss of that organization. All self-defensive expressions are ultimately futile, in the sense that all forms are ultimately transformed into new forms as energy flows through the world.

Ecclesiastes nicely expresses the futility of everything that can be done in the world:

I have seen all the works that are done under the sun,
And behold, all is vanity and a chasing after wind.

It cannot be stressed strongly enough that the only way self-defensive expressions can go forward is if the observer of a mind identifies itself with its body-based self-concept. Without its self-identification with the form of that body, it has nothing to defend.

The emotional self-identification of an observer with its body-based self-concept can only arise if that observer really feels like it is embodied within that body, as it perceives the emotional body feelings expressed by that body. Its self-identification with the form of its body is its emotional attachment to that body, but that is only a belief that it believes about itself. That belief only arises as a perception, in the sense that perceiving is believing. Its self-identification with its body is the meaning it gives to that perception.

Only a coherently organized body can express self-defensive emotions, which inherently defend the survival of the form of that body, and inherently are limited to that form. The flow of energy through a body can come into alignment with the flow of energy through other things, and create an emotional attachment to those forms, but if the body is to survive and self-replicate its form over a sequence of event, that coherent flow of energy at most can become limited to the body itself. Self-defensive expressions by a body always create a sense of self-limitation and self-identification with the form of that body.

As a body-based self-concept is emotionally constructed in a mind, and the emotional body feelings inherent in that self-concept are perceived, the observer of that mind feels self-limited to that body. Only those self-defensive emotional expressions make belief in a self-concept believable, which is a false belief that the observer of the mind believes about itself. It believes that it is embodied and self-limited within the form of that body. It believes that it is the ego, or a body-based self-image inherent in its self-concept.

Only the observer of the mind observes that body-based self-image as it arises in emotional relationship to the images of other things that appear in the world. The nature of the self-concept only arises with emotional relationships, as images are held in mental imagination and self-replicated over a sequence events. The falseness of the ego arises as the observer of the mind mistakenly identifies itself with a body-based self-image it observes. The self-concept arises in emotional relationships that relate the self-image to images of other things with emotional body feelings. The observer of the mind identifies itself with that self-image since it really feels like it is embodied in a body that expresses those emotional body feelings. In this sense, the self-concept is a false belief the observer of the mind believes about itself. As it perceives the emotional body feelings inherent in the mental construction of its self-concept, it believes its true nature is embodied within the form of a particular body it perceives on the viewing screen of its mind.

That false belief is the observer's self-identification with the form of that body. That belief is false in the sense that the viewing screen of the mind only arises from the point of view of the observer. As the viewing screen arises, a presence of consciousness arises at that central point of view. That world arises the same way a dream arises from a dreamer. That world always belongs to the dreamer. Everything in that world, including its body-based self-concept, belongs to the dreamer. It only mistakenly identifies its true nature with the central character of that dream, which is its false self-identification with the form of a particular body that appears in that world from a particular point of view. The process that deconstructs the self-concept only goes forward with the willingness of the observer to disbelieve that false belief that it believes about itself.

The focus of attention of the observer on its self-concept leads to an investment of emotional energy in its mental construction, and to an emotional attachment to that particular body. The attachment process arises as body desires are satisfied, the flow of energy through the body comes into alignment with the flow of energy through other things in the world, and feelings of connection are expressed. All attachments are limited in nature, and ultimately must become limited to the body itself. Only a coherently organized body can express emotions. Self-identification with that body arises as desires are frustrated, the flow of energy through the body goes out of alignment with the flow of energy through other things, feelings of disconnection are expressed, and expressions of self-defensiveness arise. With this expression, the observer really feels like it is embodied as it perceives those body feelings, and emotionally identifies itself with that body.

The process of awakening always begins as a process of disillusionment and discontent. Only the observer of the mind can see the falseness of its own ego. Only that self-reflective process allows the observer to detach itself from its ego. The observer is not observing its own image, but only a body-based self-image with which it identifies itself. If the ego's falseness is clearly seen, discontent arises. Discontent is the desire to destroy the falseness of the ego, which is the emotional energy that allows the ego to fight for its own self-destruction. The ego fights for its self-destruction, but that war only comes to an end with a surrender and willingness to suffer ego death rather than live the life of a lie.

The process of ego death is always a withdrawal of attention away from the ego, and a withdrawal of emotional energy in the mental construction of the ego, which is the only way the emotional construction of the ego is deconstructed. The withdrawal of emotional energy from its mental construction is a de-animation of the ego. Every expression of desire requires the expenditure of energy. Only an observer can withdraw its attention away from its ego, and withdraw its investment of energy in the emotional construction of its ego. This deconstructive process only goes forward with disbelief, if the self-concept is clearly seen as a false belief the observer believes about itself. Only if the observer clearly sees the falseness of its self-concept is it willing to deconstruct its ego, and detach itself. The de-animation of its ego is the only way the observer of a mind can detach and de-identify itself from its ego, which only goes forward with disbelief.

This self-destructive process is like the burning that occurs as an unstable state of high potential energy transitions to a more stable state, and releases heat that is radiated away. The withdrawal of emotional energy away from the mental construction of an ego is like the burning of the ego. The most stable state possible is the unchanging 'stateless state' of void. Everything ultimately burns down to nothing, just like a virtual particle-antiparticle pair that annihilates back into nothing. Those virtual pairs only appear to create a world of matter, as they appear to separate at the event horizon that is observed by the observer present at that central point of view. The nothingness of no-self is what remains as that observer detaches itself from everything in that world, enters into a state of free fall through empty space, and that entire world disappears.

He who follows the Tao
Is at one with the Tao

Returning to the source is stillness, which is the way of nature

*Stand before it and there is no beginning
Follow it and there is no end*

*The form of the formless
The image of the imageless
It is called indefinable and beyond imagination*

The farther you go, the less you know

The Tao expresses the limits of all learned knowledge, like scientific knowledge, and its ultimate limitation vis-à-vis the ultimate knowledge. All learned knowledge is the nature of imagination, and is a part of the world of images we perceive. The ultimate knowledge isn't a part of the world of images we perceive, isn't imaginary, and isn't learned. It is awareness aware of its true nature, through dissolution into its true undivided formless nature, the indescribable experience of 'knowing nothing'.

*Those who know are not learned
The learned do not know*

*In the pursuit of learning, everyday something is acquired
In the pursuit of Tao, everyday something is dropped
Less and less is done
Until non-action is achieved
When nothing is done, nothing is left undone
The world is ruled by letting things take their course
It cannot be ruled by interfering*

The correct metaphor for the deconstruction of the ego is the burning of the ego. The ego is mentally constructed as a meta-stable state of high potential energy, which is characterized by a potential barrier. In some sense, that potential barrier is the expression of self-defensiveness. The expression of self-destructiveness is like hot emotional energy that counteracts the expression of self-defensiveness, and overcomes that potential barrier. Before that meta-stable state transitions to a more stable state, heat must be applied, just as heat must be applied before the meta-stable state of a high potential energy molecule will burn. Heat must be applied before the ego will burn. The expression of self-destructiveness is like hot emotional energy that allows the ego to burn. Once the ego begins to burn, heat is released and is radiated away. The withdrawal of emotional energy away from the mental construction of the ego is like the burning of the ego. Like any other kind of burning, it begins suddenly, like an avalanche, and accelerates.

The emotional energy used to construct the ego is only withdrawn away from the ego through a process of surrender, which is the only process that deconstructs the ego. If that emotional energy is not withdrawn away from the mental construction of the ego, there is no transition to a more stable state, and there is no burning of the ego, no matter how much hot emotional energy is applied through the expression of self-destructiveness.

The process of burning away the ego only begins with an examination of the ego. That deconstructive process only goes forward if the observer of a mind looks within its mind and sees the falseness of its self-concept, from which arises the desire to destroy that false self-concept. The expression of self-destructiveness counteracts the self-defensive expressions of the ego. The expression of self-destructiveness is like the hot emotional energy that allows the ego to transition to a more stable state, but only if heat is also radiated away, which only occurs with a surrender, and the withdrawal of emotional energy away from the ego's mental construction.

The burning away of the ego is expressed by the Buddha in the Fire Sermon:

Burning, burning, burning, burning
O Lord, thou pluckest me out

Unless there is willingness to surrender, there is no forward movement. The meta-stable state of an ego cannot transition to a more stable state unless energy is radiated away, no matter how much hot emotional energy of self-destructiveness is applied to the ego. Surrender is withdrawal of emotional energy away from the ego's mental construction.

The falseness of the ego is not only seen in one's own self-concept, but can also be seen in the self-concept of others. The expression of self-destructiveness can be expressed against the falseness of a self-concept as it arises in others, just as much as it can be expressed against a false self-concept that arises in oneself. Hatred of others is a strange combination of expressions of self-defensiveness and self-destructiveness. One's own false self-concept is defended, while the false self-concept of another is more clearly seen, and is attacked with the desire to destroy that false self-concept. The problem with hatred of others is that one's own false self-concept is not seen clearly enough, while the falseness of others is seen too clearly. The biblical instruction to 'pluck the plank from your own eye' rather than remove 'the speck' from the eye of another is as valid now as it has ever been. That speck is the emotional blinders of the ego that obstructs clear seeing.

Unless there is willingness to surrender, there can be no forward movement, but only the state of being stuck with an ego. Obstructions arise as desires to hold onto things, control things, and force things to satisfy desires are expressed. Surrender is willingness to let go of things, sever attachments, and relinquish the desire to control things. It is willingness to abandon expressions of self-defensive personal will and accept universal will.

Surrender expresses the willingness to accept everything as it is every moment, with no desire that things be any different. In a state of surrender, the flow of energy through a body comes into alignment with the universal flow of all things. The nature of universal flow is to follow the path of least action, since that is the most likely path in the sense of quantum probability. All expressions of self-defensive personal will are an interference with the expression of universal will, and the normal flow of all things. The nature of that interference is what we call an interference pattern. Quantum theory describes the nature of an interference pattern as all the alternative paths that can be taken that deviate from the path of least action. Every interference takes an alternative path, which interferes with the normal flow of all things. In a state of surrender, only the path of least action is taken.

The possibility of control only arises with the expression of personal will, and is always an interference with the normal flow of all things. Expressions of personal will arise from the potentiality of things, but deviate from the path of least action, and interfere with the normal flow of all things. That personal expression of potentiality always interferes with the normal universal expression of actuality, as the path deviates from the normal path.

The expression of self-defensive personal will is always a waste of time and energy, since that expression interferes with the normal flow of all things, and does not follow the path of least action, which is the most energy efficient path to follow. That waste of time and energy is the meaning of the wasteland in the grail legend. The grail represents the achievement of the integrated state, as the flow of energy through all things comes into alignment, and unlimited feelings of connection are expressed. The wasteland represents expression of self-defensive personal will and self-identification with the form of a body. The normal flow of all things tends to follow the path of least action, since that is the most likely path, and the most energy efficient way to act. Any alternative path interferes with the normal flow of all things, and is always a waste of time and energy.

Leibniz started a philosophical firestorm in the eighteenth century when he proclaimed that we live in "the best of all possible worlds". Voltaire sarcastically responded to Leibniz with the writing of Candide, in which he demonstrated the absurdity of this claim. In retrospect, as revealed by modern theoretical physics, we can now see that Leibniz was correct, and that Voltaire was mistaken.

Leibniz made his claim based on mathematical properties of classical physics, which show that the path of any particle as it moves through space over time is to follow the path of least action. This was a brilliant insight, which Voltaire was only able to criticize since neither man knew about quantum physics and the second law of thermodynamics. If these two new ingredients are added to the argument, then Leibniz is proven to be correct. In another sense, Voltaire is also proven to be correct, since these two men were not writing about the same kind of world.

All things in the universe are animated by the flow of energy. Energy only flows through the universe as the universe evolves from more ordered to less ordered states, or as the universe evolves from its big bang event to its heat death. Every event is a decision point where the quantum state of the universe branches into alternative paths. The most likely path, in the sense of quantum probability, is the path of least action. The most likely way for anything to become animated is to follow the path of least action, which is the normal path. The path of least action is also the most energy efficient path, since it minimizes expenditure of kinetic energy while it maximizes preservation of potential energy. A world that evolves in this way is indeed the best of all possible worlds.

But that is not the way the observable world of an observer evolves with the expression of personal will, which is always an interference with the normal flow of all things. Expression of personal will deviates from the path of least action, and in that sense is a waste of time and energy. The best of all possible worlds only arises in a state of non-interference with the normal expression of universal will, which is the kind of world that Leibniz wrote about. The kind of world that Voltaire wrote about is a world characterized by ego, and the expression of personal will. These are very different worlds.

Ramesh Balsekar nicely describes this state of non-interference:

No personal individual effort can possibly lead to enlightenment. On the contrary, what is necessary is to rest helpless in beingness, knowing that we are nothing-to be in the nothingness of the no-mind state in which all conceptualizing has subsided into passive witnessing. In this state whatever happens will be not our doing but the pure universal functioning to which we have relinquished all control.

Hindu philosophy comes closest to a scientific description of the true nature of reality. The Hindu concepts of the Creator, the Preserver, and the Destroyer are inherent in any unified theory like string theory, which embodies the holographic principle of quantum gravity. Inherent in a unified theory is an empty background space within which a universe is created on a cosmic event horizon, like a bubble in the void. The nature of the void is that empty background space that physicists call the vacuum state. That empty background space, referred to in Hinduism as Paramakash, is the Absolute nature of existence, in the sense that the form of all the things that appear to exist in any world arise from that nothingness. The process of creation can only arise with quantum uncertainty, as virtual particle-antiparticle pairs appear to separate at an event horizon, as observed by the observer present at that central point of view. That apparent separation creates a holographic virtual reality, as virtual particles appear to become real, and bits of quantized information are encoded on the surface of the horizon. Those bits of information tend to align with each other due to quantum entanglement, and are spontaneously organized into coherently organized bound states of information, which is the nature of the form of all things that tend to self-replicate form over a sequence of events in the flow of energy. Self-replication of form, while behaviors are enacted, is the nature of preservation of form. As energy flows in the sense of thermodynamics, forms tend to become disorganized. Eventually all forms are destroyed. All virtual particles eventually annihilate with their antiparticles. As a world of form holographically arises on an event horizon, an observer is divided from the 'one' source of consciousness, and is present at a point of view. That observer can return to its true undivided formless state through dissolution into that nothingness, referred to as Nirvana.

Shankara refers to the absolute nature of reality as Brahman, the ultimate impersonal reality that underlies everything in the world, the source from which all things arise and to which they return. He refers to the divided presence of consciousness that perceives that world as Atman, or the Self. This is what he has to say about the nature of the world:

Brahman is the only truth. The world is illusion, and there is ultimately no difference between Brahman and Atman.

The Self refers to the presence of consciousness that perceives the entire world displayed in the mind of a person. In that world, the body of that person appears as an animated form of information, just like the animated form of a computer generated avatar in a virtual reality world. The Source of that world, and the Source of that individual presence of consciousness, is a void of undifferentiated consciousness.

Socrates tells us to *Know Thyself*, but also has this to say about the nature of death: "To fear death, my friends, is only to think ourselves wise, without being wise; for it is to think that we know what we do not know". Body death is only a transformation of form into new form. The divided consciousness of an observer can only be present for the form of a body, or return to its true undivided formless state through dissolution.

Knowing the Self is enlightenment

To die but not to perish is to be eternally present

Brings freedom from the fear of death

Eugen Herrigel describes the path of return in *Zen in the Art of Archery*:

He must dare to leap into the Origin so as to live by the Truth and in the Truth, like one who has become one with it. He must become a pupil again, a beginner; conquer the last and steepest stretch of the way, undergo new transformation. If he survives its perils then is his destiny fulfilled; face to face he beholds the unbroken Truth, the Truth beyond all truths, the formless Origin of origins, the Void which is the All; is absorbed into it and from it emerges reborn.

The Bhagavad-Gita clearly describes the path of return in the sense of detachment from things, and non-interference with the normal flow of all things:

Act, but do not reflect on the fruits of the actions.

In him whose mind dwells on the objects of sense with absorbing interest, attachment to them is formed; from attachment arises desire; from desire anger comes forth.

Poor and wretched are the souls who make the fruit of their works the object of their thoughts and actions.

Fixed in yoga do thy actions, having abandoned attachment, having become equal in failure and success; for it is equality that is meant by yoga.

The fixed and resolute intelligence is one and homogeneous.

One whose intelligence has attained to unity, casts away from him even here in this world of dualities both good doing and evil doing.

When a man expels all desires from the mind, and is satisfied in the Self by the Self, then he is called stable in intelligence.

The sages who have united their reason and will with the Divine renounce the fruit which action yield, and liberated from the bondage of birth, they reach the status beyond misery.

When thy intelligence shall cross beyond the whirl of delusion, then shall thou become indifferent.

Such of the roving senses as the mind follows, that carries away understanding, just as the winds carry away a ship at sea.

He attains peace, into whom all desires enter as waters into the sea which is ever being filled, yet ever motionless.

One who has utterly restrained the excitement of the senses by their objects, his intelligence sits firmly founded in calm self-knowledge.

Who abandons all desires and lives and acts free from longing, who has no "I" or "mine", he attains the great peace.

Fixed in that status at his end, one can attain to extinction in the Brahman.

There is no reason for the existence of the 'existent one'. There is no reason for being. Being is prior to creation and perception, prior to identification with form, and prior to whatever forms appear to exist in the world. Those forms appear to come into existence on a 'plane of existence'. The 'existent one' is the source of the light of reason, and the source of all things that appear to exist in any world. That light is divided from the darkness with the creation of that world. With the disappearance of that world, that light returns to its true undivided formless state. That primordial state of existence can be called the 'non-existent', if by existence we mean 'being something in the world'.

Being is born of not being

Being at one with the Tao is eternal
And though the body dies, the Tao will never pass away

Because there is no place for death to enter

The Buddha had something like this to say about the nature of the self-concept: 'you are what you think you are'. The conclusion of this statement is very simple. If there are no thoughts, then 'you are not'. The Tao also tells us: 'being is born of not being'. The possibility of 'being something' is created out of 'being nothing'. The only way to know that nothingness is through a process of detachment from everything that appears to exist within the world we perceive. Spiritual awakening is always a process of detachment.

Awakening is consciousness non-identified with form. Awakening can occur while forms are still perceived, or as forms disappear. Forms are perceived if there is awakening within the dream, and forms disappear if there is awakening from the dream, but in either case, consciousness is not identified with form. Expression of self-referential thought is how a presence of consciousness identifies itself with the form of its body-based self-concept. The end of thoughts is how it detaches itself from its self-concept.

Consensual reality is just like a shared dream. The consensual reality of a world is shared among many observers, as observed from many different points of view. An observer can awaken from its dream if the dream comes to an end, but can also become lucid and awaken within its dream. The emotional construction of the ego is de-animated in the lucid state, but that world isn't de-animated. In the lucid state, the mind becomes silent. The mind only becomes silent if self-referential thoughts are no longer emotionally constructed. Without self-referential thoughts, the observer of a mind no longer has a mentally constructed self-concept with which to identify itself, and knows itself only as the silent observer of that world. It knows itself to be a presence of consciousness.

Let the mind become still

Surrender is the only way the mind becomes silent, as self-referential thoughts are no longer emotionally constructed in the mind. Thoughts are only an emotional relationship constructed in the mind, as a body-based self-image is emotionally held in mental imagination and is emotionally related to the images of other things also held in mental imagination. The self-concept only arises as the observer of the mind identifies itself with that self-image. An observer of a mind only becomes aware of its own presence, and knows itself as a silent witness of its world, if its mind becomes silent. Only with mental silence can the observer know itself as a pure presence of consciousness. Surrender is the only way the flow of energy through the body comes into alignment with the flow of energy through all things in the world, as universal will is accepted. Only in a state of non-interference can the observer feel connected to all things in the world.

The lucid state is only possible with an emotional transformation, which is always a death-rebirth process. The segregated, self-defensive ego dies away, and an integrated self is reborn. The death of the segregated self is a deconstruction of form, and the rebirth of an integrated self is a transformation into new form. That emotional transformation is a change in the way form is organized, like a phase transition. Ice only melts into liquid water if a lot of heat is applied. The only way that transformation goes forward is with the expression of self-destructiveness, which is like heat that melts away the old form. The correct metaphor for the deconstruction of the ego is the burning of the ego, but in a thermodynamic sense, the burning away of the ego is not unlike melting away.

The segregated self is organized through the expression of self-defensiveness. Since its form is body-based, that form survives through self-defensive expressions, which defend body survival. Self-replication of form is what the expression of self-defensiveness is all about, and self-identification with the form of a body is what the ego is all about. The self-concept is emotionally constructed in mental imagination with self-referential thoughts that are only like a story that the mind tells about how the body survives in the world. The problem isn't the body surviving in the world. The problem is the story told in mental imagination about how a body survives in the world. The self-concept only arises with that story. That story is composed of self-referential thoughts, which are mentally constructed beliefs that relate self to other. Implicit in each belief is the false belief the observer of the mind is embodied within the body of the central character of that story.

That false belief is what the observer of the mind believes about itself as it perceives a mentally constructed belief, and feels like it is embodied as it perceives an emotional body feeling inherent in that belief. The problem isn't a body that expresses emotions in a world. The problem is a mind that emotionally constructs self-referential thoughts, which emotionally relate the concept of a body-based self to the concept of another. All those self-referential thoughts are false beliefs the observer of the mind believes about itself.

The observer of a mind only detaches itself from its self-concept if it no longer believes that false belief about itself. The only process that detaches itself from its self-concept is to sever emotional attachments. An observer is self-identified with the form of a body due to those attachments. The only way it detaches itself is to sever them. The process is straightforward. The observer looks within its mind at its self-concept and sees its falseness. As the falseness of a self-concept is clearly seen, disillusionment arises, and the desire to destroy that self-concept. Attachments are severed with a surrender and the willingness to let go, which is a process of ego death. The Gordian knot of attachment cannot be untied, only severed. That is how the war of self-destruction comes to an end, but only if battle after battle is fought, which is like a death by a thousand cuts. The observer sees the battlefield as it sees how its self-concept is constructed out of those attachments. It is only willing to fight those battles, and let go of those attachments, if it clearly sees that its self-concept is an illusion, and that it would rather suffer ego death than live the life of a lie. This self-destructive process is the 'dark night of the soul'.

This self-destructive process only goes forward if attachments are severed. That is the only way the ego is transcended. A self-concept is only constructed in the mind as a body-based self-image is emotionally related to images of other things that appear in the world. This emotional construction process is inherently self-defensive in nature, and only attachments are defended. It only makes sense to defend attachments, but the sense of 'making sense' is based on misperception. The answer isn't to 'stop making sense', but to stop making misperceptions. The ultimate attachment is to the body itself, which is the ultimate misperception. An observer attaches itself to its body as it perceives self-defensive body feelings, and feels like it is self-limited to that body. Its belief in a body-based self-concept is its self-identification with the form of that body. Since it feels like it is embodied, the survival of that body is defended as though its existence depends on it.

That observer is never really embodied or limited to the form of its body. It only believes that it is embodied due to its self-limiting beliefs. Those self-limiting beliefs are self-referential thoughts emotionally constructed in its mind, which relate its body-based self-concept to the concept of something else in its world with body feelings. That observer only believes it is embodied since it really feels like it is embodied as it perceives emotional body feelings. It only detaches itself from its false self-identification with the form of its body if it disbelieves those false beliefs about itself. The process of detachment only goes forward if it clearly sees the falseness of those beliefs, and if those beliefs are no longer emotionally constructed in its mind. The only way emotional energy is withdrawn from the mental construction of its false self-limiting beliefs is through disbelief. Withdrawal of emotional energy is always a surrender, and the willingness to abandon expressions of self-defensive personal will and accept universal will.

The emotional transformation that leads to the lucid state allows an observer to know itself as a presence of consciousness that is only present for the form of its body. It knows itself only as a witness. It knows itself as an observer present at a still point that only witnesses the form of that body, as those images play like movie images on a viewing screen. That transformation is only possible with the development of the integrated state, which is always a death-rebirth process, and a transformation of form into new form. That transformation only occurs if the observer of a mind clearly sees the falseness of its own ego, and clearly sees the falseness of all the self-defensive expressions that defend the form of its ego as though its existence depends upon it, which keeps it self-identified with that form. As it sees that falseness, self-destructiveness arises, which is the hot emotional energy that counteracts the self-defensiveness of the ego, like heat that melts ice back into water. That is the deconstructive part of the death process, but there is also a reconstructive part of the rebirth process. The segregated, self-defensive ego dies away, and the integrated self is reborn. That rebirth only occurs with a surrender. An integrated self is integrated only to the degree the flow of emotional energy through a body comes into alignment with the flow of all things. That alignment only occurs with acceptance of everything as it is every moment, with no desire to change or control anything. That acceptance expresses trust in universal will to sort out what is best for all things. The reward of that integration is unlimited feelings of connection to all things.

The focus of attention of a presence of consciousness on its world always leads to an investment of time and energy in that world. Time and energy belong to the observer as much as anything else in that world. Whatever the observer focuses its attention on in its world is what occupies the attention of the observer, and is how its time and energy is spent in that world. Every expression of desire involves the expenditure of energy. If the observer focuses its attention on distractions, that is how it spends its time and energy. If the observer focuses its attention on its mentally constructed self-concept, that is how it spends its time and energy. If the observer focuses its attention on the process of its own awakening, that is how it spends its time and energy. Only if the observer clearly sees the falseness of its self-concept will it turn away from it, withdraw the focus of its attention on it, withdraw its investment of emotional energy in it, and finally bring its mental construction to an end. That is the only way it can shift the focus of its attention away from its self-concept, and onto its own sense of being present for that world.

Why is the observer willing to surrender and detach itself from its false self-identification with its body? Why is it willing to stop interfering with the normal flow of all things, stop expressing self-defensive personal will, stop defending the survival of its body as though its existence depends on it, stop identifying itself with the form of its body, and simply watch in a state of detachment as its body is transformed into a new form? Surrender follows naturally from seeing what is. It sees the true nature of its existence. It sees that what 'it is', is nothing but undivided formless pure being, which is at the source of everything. It sees it has nothing to gain, lose, or defend. It sees it has nothing to choose. It sees that body death is only a transformation of form into new form. It sees that its true nature cannot die, but only return to its true undivided, formless state of pure being.

A presence of consciousness is willing to surrender if it identifies itself only with pure being, and not with any form it perceives. It sees the world is no more real than a dream, and any form is no more real than a character in a dream. It sees the form of everything is an illusion created out of the nothingness of pure being. It does not identify itself with any form, but only with undivided, formless pure being.

At the source of everything, there is nothing but undivided, formless pure being. The form of everything in a world is an illusion created out of the nothingness of pure being, like a dream that arises from a dreamer. As a world of form is created out of nothing, a presence of consciousness is divided from that nothingness to perceive that world. The birth, life and death of a body is only the development, self-replication and disorganization of that form. A presence of consciousness cannot die. It can only return to its true undivided, formless state of pure being.

The true nature of being can never be reduced to a form perceived in the world, anymore than the true nature of a dreamer can be reduced to a character perceived in a dream. A dreamer can identify itself with a character in its dream, but that self-identification is inherently false. The true nature of consciousness can never be reduced to something perceived within the world, and yet every observer of a world knows it exists, and knows it is aware. In physical terms, the true nature of consciousness is only describable as a focal point of perception in empty space, and the true nature of being as void.

Every body belongs to an observer, but the observer is not a body. Every mind belongs to an observer, but the observer is not a mind or a body-based self-concept. Every world belongs to an observer, but the observer is beyond that world. If it is not its body, its mind, its world or its self-concept, what is it? What is beyond all these things? Even to say it is the consciousness for these things is not quite correct. What is beyond its consciousness for all these things? What is the source of everything in its world, and the source of its consciousness? The only scientific answer we can give is the void, which is the source of everything that appears in a world, as holographically displayed upon a viewing screen, and the source of the consciousness that is present at a point of view, and perceives that world. In physical terms, the void can only be described as an empty background space. An observer can know itself as the void if it dissolves back into the void. In this sense, the void is undivided formless pure being, the true nature of what it is.

Where does the individual sense of being present, the sense of 'I am-ness', come from? The individual sense of being only arises as a world arises. As a world arises on an event horizon, like a movie of images that play upon a viewing screen, a presence of consciousness arises at a point of view, and perceives that world. That individual sense of being is already a movement in duality, as a presence of consciousness is divided from the 'one' source of consciousness with the creation of that world. The individual sense of being can be imparted to any form that appears in that world, as in the self-concept 'I am identical to the form of a body'. That self-identification with a particular body is only a perception that occurs from a particular point of view, and is only possible due to the individual sense of being that arises as a presence of consciousness arises.

Without the perception of a self-concept, there is no sense of self. A presence of consciousness that returns to its true undivided formless state through dissolution into the 'one' source of consciousness no longer has an individual identity or any sense of self. There is no individual sense of being, no sense of 'I am', and no-self in dissolution. There is only 'oneness'. Dissolution into nothingness is the nature of nonduality. The individual sense of being present for a world is always a movement in duality.

Awakening is consciousness non-identified with form. The only way awakening is possible is if there is no self-identification with form. Consciousness is always present for form, but it need not identify itself with any form. A form arises on a viewing screen, and consciousness is present at a point of view, as a world is animated like a movie of images. Every movie has its central form perceived from a central point of view. That central form is a body. The mind displays an entire world that includes the body. Organs of sensory perception in a body appear to relay information about an entire world to a brain, but all the information for that world, which includes the body, is defined upon the viewing screen. That apparent relay of information includes external sensory perceptions of the world and internal perceptions of emotional body feelings, but all of those perceptions are only a holographic appearance. A presence of consciousness perceives pain in that body if that body expresses pain, but there is only suffering if it identifies itself with that form. Suffering arises with self-identification and unwillingness to let go. Without self-identification and attachment to form, it is free to let go and become pain free. It does not feel compelled to hold onto a pain-giving thing as though its existence depends on it. It is always free to return to its true undivided formless state.

And you shall know the truth, and the truth shall set you free.

The ten thousand things rise and fall while the Self watches their return
They grow and flourish and then return to the source
Returning to the source is stillness, which is the way of nature

The Tao says the wise are impartial. Worlds come and worlds go, but the infinite potentiality of the source to create and destroy worlds is inexhaustible. Forms appear to come into existence in a world. Expression of self-concerned thought about what appears to happen in that world is a sign of ignorance, and indicates self-identification with form. Everything ultimately returns to nothingness. The only way to know that nothingness is through a process of self-destruction, which the Bhagavad-Gita expresses as:

Now I am become death, the destroyer of worlds.

The only way that return is possible is through a process of self-destruction, but to destroy one's self makes no more sense than to kill one's self for one's own good. Truth realization makes no sense. No one benefits from it. The integrated state does make sense. An integrated self benefits from the integrated state, with feelings of connection to all things, expressions of creativity, and right actions that follow from clear seeing.

The Buddha nicely summarizes what is achieved in the path of return:

Truly, I have attained nothing from total enlightenment.

There are only three possible ways to live a life in the world. Hinduism expresses the three ways to live a life with the concepts of the Creator, the Preserver, and the Destroyer. A life can be lived with the expression of creativity, self-defensiveness, or self-destructiveness. With expression of self-defensiveness, a self-concept is preserved. With the expression of creativity, the self-concept is transcended. With the expression of self-destructiveness, the self-concept is ultimately destroyed. A life lived in the world with the expression of self-defensiveness is a segregated, ego-bound life, self-identified with the form of a body. A life lived with the expression of creativity is an integrated, lucid life, which transcends all limited self-concepts. A life lived with the expression of self-destructiveness ultimately returns to the formless state of no-self, which is no life at all, and cannot be desired. The desireless state is a kind of death, and is not achievable with expression of desire. The integrated state is desirable, since it allows for expressions of creativity, unlimited feelings of connection, and right actions that follow from clear seeing. The integrated state arises with the awakening of a presence of consciousness, which is only possible if it no longer identifies itself with the form of its body.

The only possible way to live a life in the world is to have a body. Even if consciousness awakens and identifies itself only with pure being, it can only live a life in the world if it has a body, but that world is like its dream, its body is like the central character of that dream, and it is like a lucid dreamer. It does not interfere with the normal flow of things.

It simply watches as all things come and go, with no identification with any particular thing, but a general sense of identity with all things, since it knows everything in its dream belongs to itself and arises from its true nature. It is willing to let things come and go, with no desire to hold onto or control anything. Authentic desires arise in the flow of all things, as an expression of universal creativity. Inauthentic desires arise with the self-defensive expression of personal will, and are seen as false, since they lead to the mental construction of a false self-concept and self-identification with the form of a body. In the lucid state, inauthentic desires are rejected as soon as they are seen as false.

Anyone who experiences the expression of creativity knows there is no ego present while creativity is expressed. Only consciousness is present. If the ego arises, with its expression of self-concerned thoughts, that is the mental block to the expression of creativity. The ego is the obstruction. After the creative process is finished, the ego only takes false credit for that creative expression, since the ego had nothing to do with it. All great artists know that creativity has nothing to do with their egos. The sculpture is already there in the block of marble. It is only necessary to see it and remove the extraneous pieces. Creative actions follow directly from clear seeing. With creative expressions, only consciousness is present, not the ego. The expression of creativity transcends the ego. Creative expressions only arise as the flow of creative energy in a body comes into alignment with the flow of other things. The ego is a mental block in creative expressions. The ego is the obstruction that prevents clear seeing through self-identification with form and self-defensive expressions. The phony ego only arises after the expression of creativity is finished, and takes false credit for creative expressions. The ego proclaims 'look what I did', but there is no ego in that creative expression, only the expression of universal creativity. The ego itself is such a creative expression.

The expression of self-referential thoughts is inherently self-defensive in nature, since it defends the self-replication of the form of a self-concept. The only way the ego is deconstructed is with a self-destructive process. In that self-destructive process, thought is used as a weapon to destroy the self-referential thoughts of the ego. That is the only process that destroys the self-concept while the body still lives.

Self-defensive expressions arise as an expression of personal will, and interfere with the normal flow of all things. A state of surrender is always a state of non-interference. Without those self-defensive expressions, a self-image is no longer emotionally held in mental imagination and related to images of other things, and a self-concept is no longer emotionally constructed in the mind. The mind becomes silent, and the observer has no self-concept with which to identify itself. This ego-deconstructive process is the only way the observer of a mind can bring itself into the focus of its own awareness.

The desire for something is the nature of an attachment, which expresses the desire to live. The expression of that desire only comes to an end through a process of detachment. The ultimate detachment occurs as the observer withdraws its focus of attention away from its world, and withdraws its investment of energy in that world. That withdrawal of energy de-animates that world, and allows that world of form to disappear. A presence of consciousness only brings itself into the focus of its awareness with a silent mind, and only realizes its true nature as it focuses its attention upon itself as its world disappears.

A purpose in life is not the same as a reason for being. As long as one has a purpose in life, a body lives. Longevity is a consequence of having a purpose in life, but that purpose is only like a role that an actor plays in a staged production. The observer only watches that performance from its seat in the audience. If the observer's only purpose is to survive in the form of its body, then body survival is its purpose. If its purpose is the expression of creativity, then creative expression is its purpose. If its only purpose is to know the true nature of what it is, then that purpose must express itself as a process of self-destruction, and ultimately leads to either body death or ego death. The ego can fight for its own self-destruction, but that war can only come to an end with a surrender and the acceptance of ego death. That acceptance of ego death is always like the final stage of grieving. In the final stage of grieving, the death of something is accepted.

The irony is that to know the true nature of what it is, a presence of consciousness must desire nothing, since its true nature is nothing. As long as the desire for something is expressed, it has a purpose in life, and its body lives. That isn't a reason for being, only a reason for living in the form of a body. That reason for living is always the desire for something, which expresses the desire to live. There is no reason for being, since the nothingness of being is the true nature of what it is. It can only pretend to be something in the world with its reason for living. If it has a mind that constructs self-referential thoughts, it knows itself as a segregated self, self-identified with the form of its body. If its mind becomes silent, it knows itself only as a pure presence of consciousness, or as an integrated self. If it has no reason for living, its mind and its world disappear, and it knows itself as the nothingness of no-self. No-mind means no world and no-self.

Although not commonly discussed in this way, the purpose of psychoanalysis in its purest form is the achievement of the integrated state. In free association, the ego talks in an uncensored way, while the observer of the mind listens. The ego displays all its emotional conflicts, and the observer watches. Those conflicts are contradictory desires to hold onto, control, oppose, run away from, and force things to satisfy desires. Emotional conflicts inevitably arise in mental imagination, as those things that are chased after with desire inevitably turn into things that are run away from with fear or attacked with anger. Conflicts are understood in the sense of object relations, as emotional conflicts between self and other. Emotional relationships are in conflict since those desires are contradictory. The point of psychoanalysis is to expose the ego and its contradictory desires, which are seen as immature. The point isn't to satisfy contradictory desires, but to resolve the conflict. The way the conflict is resolved is called integration. The observer sees the immaturity and contradictory nature of the desires, and is willing to let go and grow up. Growing up is a process of letting go. Things are no longer seen as just good or bad. Satisfaction of desire feels good and frustration feels bad, but the desire to force things to satisfy desires only creates more bad feelings. Everything is seen as a mixture of good and bad. Acceptance of the good with the bad is called integration, and is considered the nature of emotional maturity. The psychoanalyst doesn't give a tranquilizer to mask emotional conflicts, but allows them to be exposed, and allows the observer to see them. This crisis situation is only resolved with the willingness to suffer ego death. That resolution is just like the final stage of grieving for the death of something, which is always an acceptance of death.

Zen meditation in its purest form is just like this kind of psychoanalysis. There are two essential parts of the meditation. The self-destructive part is a process of clearly seeing the falseness of the ego. As the falseness of the ego is clearly seen, the desire to destroy the self-concept arises. The surrender part is willingness to suffer ego death. The observer allows its ego to die away as the focus of its attention is withdrawn away from its self-concept, and its investment of emotional energy is withdrawn away from the ego's mental construction. Withdrawal of that self-defensive emotional energy is a surrender. There must be willingness to suffer ego death. Without that self-destructive expression and surrender, the meditation cannot go forward and reach its intended goal. The important role the Zen master plays in this process is to clearly point this out to the meditator.

Ramesh Balsekar summarizes this self-destructive process:

Concepts can at best only serve to negate one another, as one thorn is used to remove another, and then is thrown away. Words and language deal only with concepts, and cannot approach Reality.

The goal of this process is the achievement of a silent mind. Only with a silent mind can the observer of the mind bring itself into the focus of its own awareness, and know itself as a pure presence of consciousness. That observer only realizes its true formless nature if the observer focuses its attention upon itself as its world of form disappears.

Let the mind become still

Be still and know that I am God

The Tao-Te-Ching literally means 'the way and its energy'. The word Tao means the way, and refers to the natural path all things take, which is the path of least action. The word Te refers to energy, which in the ying-yang symbol is understood to come in both positive and negative aspects, and to be in a state of balance, just like a virtual particle-antiparticle pair that is created out of nothing and annihilates back into nothing. That nothingness is understood as the nature of oneness, the source of everything, and the ground of being. The Tao refers to the path of return as a return to that nothingness. The Tao also describes the path of return as a state of non-interference and non-doing.

The word repent means 'to turn away', just as exodus means 'the way out', yoga means union, and atone means 'being at one'. All of these words refer to the path of return. An observer of a world can only know itself as a pure presence of consciousness if it turns away from that world and focuses its attention on its own sense of being. In this sense, the only sin to regret is ignorance, which is an observer that does not know its true nature, and mistakenly identifies itself with an image, or the form of something that it observes. Self-identification with form is the original sin. Prior to self-identification with form, an observer lives in a state of innocence, like a very young child prior to the development of a self-concept. After the self-concept develops, the observer lives in a state of ignorance, self-identified with the form of a body. Only with destruction of its self-concept can the observer live in a state of wisdom, referred to as eternal life. Only a body can die as form becomes disorganized. An observer can only return to its true state of timeless being.

That timeless state of being is not a thing in the world. A detached observer continues to observe its world, but if its mind becomes silent, it no longer identifies itself with any form in that world since it no longer has a body-based self-concept with which to identify itself. It shifts the focus of its attention onto itself, and knows itself only as a silent observer of its world. The Book of Psalms describes this silent state of mind as 'Be still and know that I am God'. The New Testament also describes the unchanging primordial nature of existence with the expression 'Before Abraham was, I am'.

The Bhagavad-Gita expresses the unchanging primordial nature of existence as:

Never the spirit was born
The spirit shall cease to be never
Never was time it was not
End and beginning are dreams

The Bhagavad-Gita also describes the enlightened state:

In the knowledge of the Atman
Which is a dark night to the ignorant
The recollected mind is fully awake and aware
The ignorant are awake in their sense life
Which is darkness to the sage

The Bhagavad-Gita also describes the lucid state of a detached observer:

The soul that with a strong and constant calm
Takes sorrow and takes joy indifferently
Lives in the life undying
That which is can never cease to be
That which is not cannot exist
To see this truth of both
Is theirs who part essence from accident
Substance from shadow

The lucid state is sometimes described with the mythological image of the central mountain of the world. It is as though a lucid observer looks down at its world from the central mountain of its world, which is at the center of its world. It looks down on its world, just like it looks at images on a distant horizon. It looks down on all the characters in that world, including the central character. This way of seeing the world is described as seeing things in a spiritual way. The realization of this way of seeing is non-identification with form. Every observer is at the center of its own world, and looks down on that world. That center is everywhere, since it is only another point of view in empty space. In this sense, everyone who observes a world is at the center of their own world. A lucid observer knows itself only as a pure presence of consciousness at a still point, and is not identified with anything it observes in that world. Every point of view in empty space is at the center of a potential world that arises on an event horizon. As an observer arises at that central point of view, its world arises on a distant horizon. That center is everywhere.

The lucid state is often described like the experience of a lucid dream. In a lucid dream, the observer of that dream looks down on all the characters in that dream, which include the central character of the dream. That lucid observer has the sense of being 'in the dream' as the central character, but also has the sense of not being 'of the dream'. That observer knows it is not a character in its dream. It is nothing it perceives in its dream. It looks down on its dream, and it sees how the central character acts, but it has the strange sense of not being self-identified with the central character, and being outside its dream. That dream is only like a performance, and the central character is like an actor on a stage. As the lucid dreamer perceives that performance, it knows itself to be outside that stage, in the audience. It is as though the performance plays like a movie of images on a viewing screen while it observes the performance from its seat in the audience. The lucid dreamer does not control what appears to happen on the stage. It only watches the performance, which arises in the normal flow of all things. It is not 'of the dream', since it is not identified with the central character. There is another sense in which it is not 'of the dream'. It can awaken from its dream. If it awakens, the dream disappears.

The lucid state is often described as the theater of consciousness in our greatest literature. Shakespeare refers to the world as a stage, populated by actors on the stage:

All the world's a stage
And all the men and women merely players

Who is out there in the audience of the theater of consciousness, watching this play? To paraphrase Gertrude Stein: Is there really any 'there' there? Shakespeare does not give an answer, but does describe the futility of everything that can be done in the world:

Life is but a walking shadow, a poor player
That struts and frets his hour upon the stage
And then is heard no more. It is a tale
Told by an idiot, full of sound and fury
Signifying nothing

Plato also refers to shadows as he describes the theater of consciousness and the lucid state. In the Republic he gives it emphasis in the section called the Allegory of the Cave:

They see only their own shadows, or the shadows of one another, which the fire throws on the opposite wall of the cave.

To them, the truth would be literally nothing but the shadows of the images.

See what will naturally follow if the prisoners are released and disabused of their error.

See the reality of which in his former state he had seen the shadows; and then conceive someone saying to him, that what he saw before was an illusion.

His eye is turned towards more real existence, he has a clearer vision.

Shakespeare refers to shadows the same way Plato describes the shadows of images displayed on the wall of the cave, just like animated images displayed in a movie. Plato describes prisoners who observe the shadows, and mistake those shadows for their true nature. The prisoners believe something about themselves that is untrue. The prisoners believe they are the shadows they perceive. In a sense, the prisoners only believe that false belief about themselves since that is the way it 'feels' to them as they perceive it, and feeling is believing. Perceiving is believing, which is another way to say 'to act is to give meaning'. The meaning that we give to the things we perceive in the world arises with emotional actions, which we perceive as body feelings. Only emotional expressions make beliefs believable, and are inherent in all beliefs. Belief is not possible without a body.

Beliefs only come to an end with the end of the emotional expressions that make those beliefs believable, which is the end of belief in a body-based self-concept. Socrates expressed this idea with his famous saying *Know thyself*. This is the motto for the movie the Matrix, which is a retelling of the Allegory. The Matrix is about a virtual reality created within the theater of consciousness. The story is about a prisoner self-identified with a character in that virtual reality, the journey that allows for escape from that prison, and the end of that false belief. That journey allows a knower to know its true nature, and no longer believe it is something it perceives. That knower believes it is the animated form of an image it perceives since that is the way it really feels, and feeling is believing.

The central character of the Matrix is told 'you are the one', but is also told 'you've been living in a dream world', and when given the chance to awaken is told 'all I'm offering is the truth, nothing more'. Shakespeare also tells us that 'life is but a dream':

Our revels now are ended. These our actors,
As I foretold you, were all spirits and
Are melted into air, into thin air:
And, like the baseless fabric of this vision,
The cloud-capped towers, the gorgeous palaces,
The solemn temples, the great globe itself,
Ye all which it inherit, shall dissolve

And, like this insubstantial pageant faded,
Leave not a rack behind. We are such stuff
As dreams are made on, and our little life
Is rounded with a sleep.

In this quote Shakespeare refers to actors on the stage (the characters), the spirits (the presence of consciousness), the dream-like illusion of the world (the baseless fabric of this vision), and the disappearance of the world (the great globe itself-shall dissolve). It makes one wonder if Shakespeare was enlightened.

To sleep: perchance to dream: ay, there's the rub;
For in that sleep of death what dreams may come
When we have shuffled off this mortal coil
Must give us pause

Thoughts are actions like anything else that one can do, except they are acted out in mental imagination. When one does nothing, one also thinks nothing. Doing nothing includes thinking nothing. The focus of attention of consciousness on anything leads to an investment of emotional energy in that thing. Only that flow of energy allows for animation of that thing over a sequence of events. Thinking is a symbolic representation of doing constructed in mental imagination, and requires the same investment of emotional energy. With thinking, the focus of attention is on a body-based self-concept held in mental imagination, as that self-concept is emotionally related to the concept of other things in the world. If the focus of attention is withdrawn from the self-concept, self-referential thoughts are de-animated, and the observer of the mind thinks nothing. Without self-referential thoughts, the observer has no self-concept with which to identify itself, and knows itself only as a pure presence of consciousness, or a silent witness of that world. It witnesses its world from the immovable center of that world, while all the images of that world play like a movie of images on a distant screen.

Events still happen in the world, but all things tend to follow the path of least action in the normal flow of things. That is the normal way for the world to be animated. The detached observer does not interfere with the normal flow of things, but simple watches as things appear to happen. A detached observer does nothing in its world, but only looks down on that world like images that play on a distant horizon. The body of the central character is only another animated image on the screen. The detached observer is not self-identified with the central character, and no longer feels embodied in that body.

With the lucid state, the mind becomes silent with the acceptance of everything as it is, without the desire to explain anything. Explanations are seen as false, since all explanations are inherently self-defensive in nature. Ultimately, nothing can be explained. As Einstein pointed out, everything that appears to happen in a world is probabilistic in nature, and is no more explainable than a dice game. Explanations only arise in that world like the narration of an animation by the central character of that animation. The end of self-referential thoughts is like the end of the narration of that animation by the central character. The lucid dreamer continues to observe that world, but knows it is nothing it observes, and is only witnessing that world.

The de-animation of the ego is only possible in a state of non-interference. That world is still animated and actions occur in that world, but without self-defensive expressions of personal will and an ego that obstructs clear seeing, the lucid dreamer clearly sees the course of right action, which is the path of least action. Right action follows from clear seeing. There is simply acceptance of everything as it is every moment, with no desire to change or control anything. Things are accepted as they come and go, with no desire to hold onto things, and no desire to oppose anything. Actions arise naturally in the universal flow of all things without expressions of personal desire. With that acceptance, the flow of energy through the body comes into alignment with the universal flow of energy through all things, and feelings of connection to all things are expressed. Unlike emotional attachments, unlimited feelings of connection only arise in the detached state.

A detached observer that enters into a lucid state feels connected to everything that appears in that world. That lucid observer knows that all things come and go like clouds in an empty sky, and knows itself to be a presence of consciousness that is present at a still point in that empty space. That detached observer is present for that movement, but is not attached to anything that appears to move and change. That detached observer knows that everything appears to move through expression of universal will, and knows itself only as a pure presence of consciousness that only observes that motion and change. It knows death is only a transformation of form into new form. It knows it can only be present for the form of things, or return to its true undivided, formless state of pure being.

The lucid observer knows that the expression of universal will is a manifestation of universal intelligence, since consciousness is present for all things. The lucid observer puts its trust in universal intelligence to sort out what is best, and identifies itself only with the 'one' source of consciousness. It remains detached, but in a friendly way, without any desire to attach itself to anything that appears to change in the world. That observer simply witnesses things as they come and go in the flow of energy through all things. There is a normal curiosity about that world, and a sense of amusement and interest in things, but that observer remains unattached to anything that arises in that world. But the final detachment from the world only occurs with the de-animation of that world.

The only way to understand detachment from everything, and the ultimate state of free fall through empty space, is with the concept of a force. The principle of equivalence tells us every force is equivalent to an acceleration. An acceleration implies an accelerating frame of reference with an observer present at the central point of view, which we can take as the origin of that frame of reference. Every accelerating frame of reference has an event horizon. A horizon is as far in space as an observer can see things in space, due to constancy of the speed of light. The holographic principle tells us all the information for all the things observed in that space is encoded on the event horizon, which acts as a viewing screen that projects images to the central point of view. The motion of all things in that space is only a holographic appearance, and that motion is relative, as all things appear to move relative to each other. It helps to deconstruct the principle of equivalence all the way down to an empty space that is only characterized by the information encoded on the viewing screen and the observer at the central point of view.

When a person stands on the surface of the earth, that person experiences a gravitational field. All things dropped above the surface of the earth fall down with an acceleration rate a=g relative to the person. The earth's gravitational field is always equivalent to a spaceship that accelerates through empty space with an acceleration rate a=g. A person that stands on the floor of an accelerating space ship experiences the same acceleration of things dropped above the floor of the spaceship, which fall down with an acceleration rate a=g relative to the person. But that person is not itself in a state of free fall. Due to quantum uncertainty, the atoms in the body of the person cannot occupy the same space as the atoms in the floor of the spaceship, or the atoms that compose the surface of the earth. There is an effective repulsive force between the body and that surface. The floor of the spaceship, or the surface of the earth, holds up a person that stands upon that surface. Even if that body is in free fall in the earth's gravitational field, we have only eliminated the effects of gravity, and not the other fundamental forces.

If we look closer inside the body of the person, we see all the atoms in that body are bound together due to the electromagnetic force of attraction between atoms. That force arises due to an uneven distribution of electric charges in space, and the electromagnetic force of attraction between negatively charged electrons and positively charged atomic nuclei. Even the atomic nuclei are bound together due to the nuclear force of attraction between quarks. In the same way, the atoms in the earth or in the spaceship are also bound together. A body of a person is just as much a bound state as is the earth or the spaceship. Unified theories unify the electromagnetic force and nuclear forces with gravity through compactification of extra dimensions. The principle of equivalence applies to those forces just as much as it does to gravity. Every force is equivalent to an accelerating frame of reference in empty space. A state of free fall through empty space eliminates the effects of all forces. But the only way to understand that state of free fall is with the viewing screen description. All forces are equivalent to an accelerating frame of reference in empty space, which always has an event horizon, where all the information for all the things observed in that space is holographically encoded.

The images of all the things observed in that space are observed from the central point of view of that frame of reference. A force is always equivalent to an accelerating frame of reference, which only arises with the expenditure of energy. Without that expenditure of energy, there is only a state of free fall. Only the observer at that central point of view can enter into a state of free fall through empty space, in which case the effects of all forces are eliminated, and all those images disappear. There is no event horizon for an observer in a state of free fall. The form of everything disappears for an observer in a state of free fall through empty space. This is a natural consequence of the equivalence principle.

The form of a body can appear to move in the world as the body moves relative to the form of other things in the world, but that appearance of movement is only an illusion, or a holographic appearance, as forms appear to move on a viewing screen. Even the bits of information on the screen do not move around in space, but only become organized into the form of things, as they come into and go out of alignment with each other. As forms are organized and self-replicated in form, they appear to move relative to each other.

Forms only appear due to the organization of information, and they disappear as information is disorganized. The observer of a world appears to move around in that world, but that appearance of movement is only another illusion created as the observer identifies itself with the form of a body with its emotional attachment to that body. The observer only feels like it is embodied in a body that appears to move around, as it perceives the emotional body feelings expressed by that body. The observer does not really move. Only a perceivable world of form is animated. If the observer no longer feels embodied within the form of a body, and no longer identifies itself with that body, then the observer knows itself only as a pure presence of consciousness at a still point, while its world plays like a movie of images on a screen on a distant horizon.

Self-identification with the form of a body is only possible as a presence of consciousness arises at a point of view, while a world of form holographically arises on a viewing screen. As the form of its body is emotionally self-replicated, the observer perceives the emotional body feelings expressed by that body, and feels like it is embodied within that body. With every expression of desire, there is expenditure of energy and the animation of a body. As an observer arises at a point of view, its sense of being present, its sense of 'I am-ness', also arises. That sense of being can be emotionally imparted to the perceived form of its body, as in the self-concept 'I am identical to the form of a body'. But that self-identification with the form of a body is inherently false, since the sense of being does not arise from the perceived form of a body. That sense of being is inherent to the observer itself. The sense of being arises as the observer arises from the 'one' source of consciousness, or the primordial nature of existence. That individual sense of being is only emotionally imparted by the observer to the form of a body since the observer really feels like it is embodied as it perceive the emotional body feelings expressed by a body. The true nature of the observer is the consciousness present at a point of view in empty space, not the perceivable form of an image animated on a viewing screen that the observer perceives. The form of its body is only another animated image it perceives.

Science only describes the perceivable world in terms of how information is encoded in that world, and how energy flows through that world. That scientific description is inherently mathematical in nature. The encoding of information and the flow of energy are described by computational rules. The universal flow of energy is described by thermodynamics, and describes how states of information evolve into other states of information. That universal flow of energy begins with a big bang event, and ends with the heat death of the universe. The encoding of information is described by quantum theory, which describes the discrete nature of how quantized bits of information are encoded in any state of information. Relativity theory requires those states of information are encoded on the surface of an event horizon, as observed by the observer present at the central point of view. Each fundamental pixel defined on the viewing screen encodes a quantized bit of information. The principle of equivalence expresses the equivalence of all observers present at all points of view in empty space. The fundamental principle of quantum theory is the uncertainty principle, which describes how virtual particle-antiparticle pairs spontaneously arise within empty space. The encoding of information on a horizon is inherently linked to the separation of virtual pairs at the event horizon.

There is something truly remarkable about the way information is encoded on an event horizon. If we look at how classical physics describes the motion of point particles, those particles follow a path through space over time, and that path is determined by a principle of least action. Since that path is continuous, there is no discrete encoding of information as in quantum theory. Thermodynamics has a hard time even defining information in classical physics due to the lack of the discrete encoding of quantized bits of information. Quantum theory solves this problem by specifying quantized states of information where information is only defined in terms of quantized bits of information, but the price of quantization of information is to require a sum over all possible paths to define a quantum state of potentiality. Thermodynamics then describes in a very natural way how one state of information evolves into another state of information. Thanks to quantum theory, all information is encoded in a discrete way in terms of quantized bits of information. The problem is how to unify quantum theory with relativity theory.

In relativity theory, there is no pre-existing space and time for point particles to move through, and to follow a path through space over time. All unified theories assume the existence of an empty background space, which is called the void or the vacuum state. The motion of a point particle through space over time is only a holographic appearance. All information is encoded on an event horizon, with one quantized bit of information encoded per pixel defined on the viewing screen. The event horizon always arises from the point of view of the observer present at the central point of view. The encoding of bits of information is holographically equivalent to separation of virtual particle-antiparticle pairs at the event horizon, as observed by the observer at the central point of view.

There is nothing objectively real about the third dimension. The concept of the third dimension is a mental concept like any other concept. That concept only arises with the self-concept, which is a self-identification with form. Those forms are inherently two dimensional images animated on a surface, which is an event horizon that is observed from a point of view in empty space. Self-identification with form creates the illusion of the third dimension. Without that self-identification with form, those images are seen two dimensionally. The concept of the third dimension only arises with the self-concept, and with self-identification with the form of a body.

An interesting fact about young children is that prior to the development of conceptual thought and a self-concept around the age of two years old, a child has no concept of the third dimension (private communication Jonathan Dickau). The world is perceived with only two dimensions for a child younger than around two years of age. A self-concept only develops in a child around the age of two with the development of conceptual thought, which only arises with the development of coherent organization. Prior to that age, a young child has no concept of self, and no concept of a third dimension. Only with the development of a self-concept can the consciousness present for the mind of the child see its world three dimensionally. Prior to this age, a change in distance to an object is not distinguishable from a change in the size of the object. For a child without a self-concept, a change in the distance to an object is only perceived as a change in the size of the object. The concept of the third dimension only arises in a very young child with the emotional development of conceptual thought and a self-concept.

This state of affairs is exactly how Susskind describes the holographic principle: "Why would a world with only two dimensions be exactly the same as one with three dimensions?" "If one projected" an image "onto the boundary by creating a shadow, the image would shrink and grow as the object approached and receded from the boundary." "From the point of view of the three dimensional interior, this is an illusion." "Growing and shrinking in the Flatland half of the duality is exactly the same as moving back and forth along the third direction in the other half of the duality." "Everything that takes place in the interior" "is a hologram, an image of reality coded on a distant two dimensional surface" (Susskind 2008, 417). Those three dimensional images are only an illusion, like the shadows projected onto a screen that is perceived at a point of view. The use of the word 'shadows' to describe the nature of that three dimensional illusion is eerily familiar to Plato's description in the Allegory of the Cave.

The world only appears three dimensional since the fundamental nature of a presence of consciousness is to be present at a focal point of perception in empty space. All the bits of information for that world are holographically encoded on an event horizon that arises in empty space, as observed by the observer at the central point of view. The world appears three dimensional since it is only a holographic projection of images from the viewing screen to the observer at the central point of view. Inherent in the perception of the third dimension is the self-identification of the observer with the form of a body perceived in that world. Without that self-identification with form, the world appears two dimensional. The perception of dimensions from a focal point of perception expresses the principle of equivalence, while encoding of information on an event horizon expresses the uncertainty principle, as virtual particle-antiparticle pairs appear to separate at the horizon.

As long as the equivalence and uncertainty principles are valid, the perceivable world can only appear two or three dimensional, due to its holographic nature. The perceivable world tells us nothing about the nature of the observer of that world. The true nature of the observer can only be described as the void, or the empty background space the world is created within. As all the images for that world arise upon an event horizon that acts as a viewing screen, and those images are holographically projected to a focal point of perception, an observer arises at that point of view, and perceives those images. The observer at that point of view is a presence of consciousness that arises from the 'one' source of consciousness, as a world of form holographically arises from the void.

The scientific ideas presented here about the ultimate nature of reality are fundamentally simple. This approach begins with the most fundamental principles of science. Through a straightforward process of logical deduction, it deduces the ultimate nature of reality. The fundamental scientific principles are simple, and the process of logical deduction is straightforward. There is really nothing very complicated about it. These ideas attempt to scientifically discuss that which cannot be simpler. The approach presented here is scientific, but leads to simplicity. Simply stated, nothing is simpler than nothing. Ultimately, there is nothing to learn, nothing to know, nothing to do, and nothing to become. At its most fundamental level, ultimate reality is that nothingness.

The key thing about this kind of scientific explanation is that anyone can think it out for oneself. Once the fundamental principles are understood, anyone can think through the process of logical deduction. This is the kind of process students go through when they prove the Pythagorean theorem for themselves. Given the fundamental assumptions of geometry, the proof is straightforward. Anyone who engages in this process in a serious way can discover the answers for oneself.

This is exactly the same kind of argument that leads to a natural explanation of spiritual enlightenment. The argument is simple, and is only based on fundamental assumptions. The key thing is to examine the assumptions, and be clear about what is assumed. Once the assumptions are clear, the process of logical deduction is straightforward. There is nothing mysterious about it. Anyone can think through the argument for oneself. This is exactly the same process of logical deduction that Plato used in his arguments.

Although Plato did not say it exactly like this, his message was very clear. Everyone can examine one's own assumptions and think for oneself. The greatest spiritual message anyone can ever be given is to think for oneself. It is always a mistake to rely on outside authority figures. Since everyone has the same access to the source of reason, it is always possible to examine one's assumptions and think for oneself.

Why is it a spiritual process to think for oneself? The most profound spiritual question anyone can ever ask oneself is 'Who am I?' Spiritual enlightenment is inevitable if this question is relentlessly followed to its inevitable conclusion. The only way a process of self-inquiry can fail to result in awakening is if one fails to do it. The question must be approached in the manner of an attack on the question itself, without compromise until all possible answers are exhausted and defeated. The self-concept only arises with answers to this question. The destruction of all possible answers is a process of self-annihilation. This is the only process that defeats the supremely confident ego and destroys character. That is the war the ego fights with itself, which only comes to an end with surrender and acceptance of ego death. Without self-referential thoughts, there is no answer. Without thoughts and a self-concept, the only possible answer is 'I am not'. The final answer is an answerless answer, the direct experience of knowing nothing and being nothing. Anyone who goes through this process knows nothing about oneself except 'I am'. Everything else one knows about is no more real than an illusory image one perceives in one's dream.

The only thing that is really sacred in this world is the awakening process itself, which is the same as to say 'nothing is sacred', or to give 'the whole truth and nothing but the truth'.

These scientific ideas about the awakening process are no more a process of awakening than reading a map is the same as making a journey to a far away and unknown land. An explanation of spiritual enlightenment is as false as any other mental concept. These scientific concepts are only like a map that points out travel directions, but anyone who is determined to make such a journey can always use a good map. The farthest that anyone ever can go is the destination called terra incognita, which is a state of unknowing. To paraphrase the Tao: 'Those that know, know nothing', and arrive at a state of unknowing.

Susskind nicely, although unintentionally, summarizes that journey: "Very likely, we are still confused beginners with very wrong mental pictures, and ultimate reality remains far beyond our grasp. The old cartographer's term *terra incognita* comes to mind. The more we discover, the less we seem to know" (Susskind 2008, 441).

Anyone who goes through the awakening process begins to see things clearly. One sees things clearly as they are every moment, without any desire that things be any different. Anyone can see things in this spiritual way if one awakens. There is nothing special about this way of seeing things. Everyone has the same ability to see things. One only has to open one's eye, or remove the emotional blinders of one's own ego that distorts one's ability to see things clearly. That distortion is one's emotional self-identification with the form of some thing one perceives. One does not just see a thing, but one sees oneself in the thing, which distorts one's ability to see the thing clearly, as one identifies oneself with the thing. This kind of emotional distortion arises with wishful thinking, mental imagination, and emotional projections driven by expressions of fear and desire.

An awakened one does not identify oneself with anything one sees. All the things one sees are only images one perceives, like movie images on a screen. One knows one is always outside those images, only present at a focal point of perception in the audience. One is external to all the things one perceives. Nothing one perceives is internal to one's true nature, and yet everything arises within one's true nature, since that true nature is nothingness, and everything perceivable arises within nothingness. One is never inside the form of something one perceives, and one does not identify oneself with anything one perceives. One can only describe what one sees, but even that description is external.

A scientific description can usefully and accurately represent what the awakened see. Science is useful to the degree it accurately represents the things we see. Science is a description of our observations. If science accurately represents the things we observe, then science is a good description of those observations. Science is only useful to the degree it conforms to our observation of things. What science can never do is give a complete explanation of the true nature of what is observing those things.

The true nature of consciousness is only describable as a point of view in empty space, and the true nature of being as void. There may be a reason for living, but there is absolutely no reason for being. It is not possible to explain how this is possible. It is what it is, or as the Book of Exodus says 'I am that I am'. It is the nature of nothingness. It is possible to explain how something is created from nothing, but only if that nothingness is the nature of consciousness. A perceivable world of form can holographically arise on an event horizon, as observed by the observer at the central point of view, since virtual pairs can appear to separate at the horizon as information is encoded on the horizon, but this is only possible if that empty background space is the true nature of consciousness.

Shankara describes the undivided, formless, non-identified nature of consciousness as that unchanging, limitless, infinite empty background space: "That which permeates all, which nothing transcends and which, like the universal space around us, fills everything completely from within and without, that Supreme non-dual Brahman – that thou art".

How is it possible for an observer to know itself at a still point? That observer is a divided presence of consciousness that arises at a point of view in empty space as a world arises on the surface of an event horizon. All the forms of information that appear in that world, like a body, are animated upon that surface. Greene describes this as: "Since there is no difference between an accelerated vantage point *without* a gravitational field and a non-accelerated vantage point *with* a gravitational field, we can invoke the latter perspective and declare that *all observers, regardless of their state of motion, may proclaim that they are stationary and 'the rest of the world is moving by them', so long as they include a suitable gravitational field in the description of their own surroundings*" (Greene 1999, 61). There are only three possibilities:

1. An observer that is emotionally attached to its body and self-identified with the form of that body. That observer appears to move in the world. That appearance of movement is an illusion created by self-identification with the form of a body.
2. An observer that becomes emotionally detached from its body and is no longer self-identified with that form through a process of de-animation of its ego. That observer knows itself only as a pure presence of consciousness at a still point.
3. A detached observer that is non-identified with form, and that enters into a state of free fall through empty space through a process of de-animation of that world. That world disappears, the effects of all forces disappear, all forms disappear, and that observer dissolves back into empty space, which is its true undivided formless state of pure being.

Only an observer that detaches itself from its world can realize its true nature. That observer only detaches itself from that world, and enters into a state of free fall through empty space, if it no longer believes that it is anything in that world. That belief only comes to an end if it is no longer believable, which only occurs through a process of emotional detachment from everything in that world. Only emotional expressions make beliefs believable, since feeling is believing. In the desireless state, belief comes to an end. All self-concepts are false beliefs that come to an end in the desireless state. What appears to happen in the world is only personal if one believes one is a person. Ultimately, an observer that clearly sees the falseness of its self-concept, as emotionally constructed in its mind, is willing to suffer ego death rather than live the life of a lie. That detached observer is no longer identified with form. All animated forms of information disappear as the observer enters into a state of free fall through empty space, and its world disappears. A detached observer has nothing to reference its fall relative to, which is its dissolution into nothingness and oneness, and its return to its true undivided, formless state of pure being.

Returning is the motion of the Tao

It returns to nothingness

It leads all things back
Toward the great oneness

There is a scene in the Matrix that expresses the incredible, almost fantastic nature of the world, which truly seems to be beyond belief. After Neo escapes from the Matrix, he returns with Morpheus and Trinity to see the Oracle. As they ride together, and Neo looks upon the virtual reality of the Matrix with awe, Morpheus says "Unbelievable, isn't it".

References

Balsekar, Ramesh. 2004. *The Bhagavad-Gita*. Mumbai: Zen Publications.

Banks, Tom. 2011. http://blogs.discovermagazine.com/cosmicvariance/2011/10/24/guest-post-tom-banks-contra-eternal-inflation-2/

Bousso, Raphael. 2002. "The Holographic Principle". arXiv:hep-th/0203101v2.

Damasio, Antonio. 1999. *The Feeling of What Happens*. New York: Harcourt.

Davies, Paul. 1977. *The Physics of Time Asymmetry*. Berkeley: University of California.

Ellis, Robert. 2011. "Taking the 'Meta' out of Physics".

Feynman, Richard. 1963. *Feynman Lectures on Physics*. Reading: Addison-Wesley.

Goldstein, Rebecca. 2005. *Incompleteness*. New York: Norton.

Greene, Brian. 1999. *The Elegant Universe*. New York: Vintage Books.

Greene, Brian. 2011. *The Hidden Reality*. New York: Vintage Books.

Lao Tsu. 1997. *Tao Te Ching*. Gia-Fu Feng, Jane English trans. New York: Vintage.

Penrose, Roger. 1999. *The Large, the Small and the Human Mind*. Cambridge: Cambridge University Press.

Penrose, Roger. 2005. *The Road to Reality*. New York: Knopf.

Plato. 1991. *The Republic*. Benjamin Jowett trans. New York: Vintage Books.

Shimony, Abner. 2009. "Bell's Theorem". Stanford Encyclopedia of Philosophy.

Smolin, Lee. 2001. *Three Roads to Quantum Gravity*. New York: Basic Books.

Susskind, Leonard. 1994. "The World as a Hologram". arXiv:hep-th/9409089v2.

Susskind, Leonard. 2008. *The Black Hole War*. New York: Back Bay.

Zee, A. 2003. *Quantum Field Theory in a Nutshell*. Princeton: Princeton University.

News

Searching for Earth's Twin

Philip E. Gibbs[*]

Abstract

This is news about Kepler space telescope's latest findings adapted from viXra log (http://blog.vixra.org).

Key Words: Earch's twin, Kepler space telescope, search, exoplanet.

December 22, 2011: Searching for Earth's Twin

This has been a great year for experimental big science with ground-breaking findings in particle physics and astronomy. One of the most remarkable breakthroughs has been the success in the search for planets around other stars. The word exoplanet which first appeared in print in 1995 according to google, and has become a popular term in news reports in just the last three, has become ever more familiar this year as reports from the Kepler space telescope have taken the number of candidate exoplanets into the thousands.

Kepler is constantly watching 145,000 main sequence stars in our nearby region of the Milky Way galaxy in the direction of the constellations Lyra and Cygnus. It is looking for the tiny dimming of light that tells us that a planet has passed in front of the stars disk. By recording the amount of dimming, how long it lasts for and how frequently it repeats, Kepler can estimate the size and orbit of the planet. In February NASA released a catalog of 1236

---

[*] Correspondence: Philip E. Gibbs, Ph.D., Independent Researcher, UK. E-Mail: phil@royalgenes.com
Note: This report is adopted from http://blog.vixra.org/2011/12/22/searching-for-earths-twin/

candidate exoplanets and this month the number increased to 2326. These have to be verified by ground based observation and so far the catalog of confirmed exoplanets has 716 entries.

The real interest about exoplanets concerns whether or not there is other life in the universe, and if there is, how common it is. A whole new industry of exoplanetary statistics has been born with scientists inventing habitability indexes that can be applied to the exoplanet catalogs to gauge which ones could support life. One habitable planet catalog has two exoplanets regarded as more Earth-like than Mars. These are HD 85512b and Gleise 581d, both found earlier this year, but they are rather large to be comfortable for us to live on. If they have an atmosphere it is could be too thick due to the stronger gravity. Already Venus has a thick atmosphere making the pressure too hot and high pressure for us to survive. If we discovered an exoplanet like Venus we would be very excited because it is in the habitable zone and is very similar to Earth in size. Finding out about its atmosphere would be difficult from a distance of many light-years.

This week some new "Earth-twins" were announced Kepler 20f and Kepler 20e. They are very similar in size to earth but they are not in their Suns habitable zone where the temperature would be about right for liquid water and conditions similar to Earth. It is good news that Kepler has proven that it can find planets of this size but we will need to wait before we find ones where we could really live. It is said that these planets may have been further from their star in the past so that life could have formed there in the past. This just serves to emphasize one more characteristic an exoplanet must have if it is likely to support life as we know it on Earth. It must stay in a stable orbit around a stable star for billions of years so that life can evolve without being obliterated by heat or freezing.

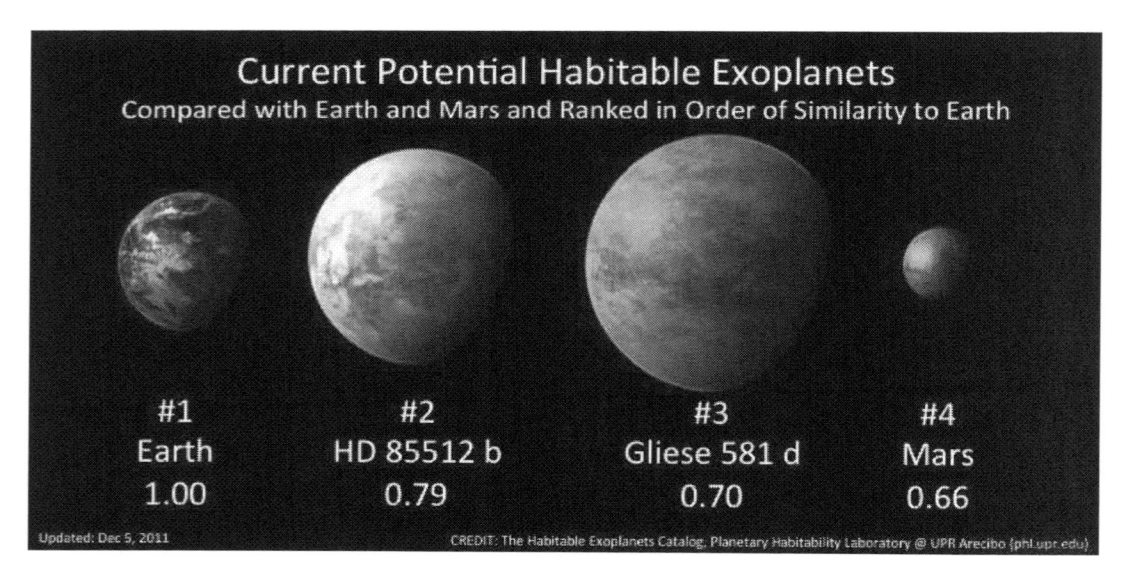

Current Potential Habitable Exoplanets
Compared with Earth and Mars and Ranked in Order of Similarity to Earth

| #1 Earth 1.00 | #2 HD 85512 b 0.79 | #3 Gliese 581 d 0.70 | #4 Mars 0.66 |

Updated: Dec 5, 2011 — CREDIT: The Habitable Exoplanets Catalog, Planetary Habitability Laboratory @ UPR Arecibo (phl.upr.edu)

Kepler has a planned lifespan of 3.5 years and may have its life extended. This should give it time to find some more earth-like planets orbiting Sun-like stars with periods of about 1 year. Kepler takes time to find these because they need to pass in front of their star at least twice to confirm their existence and orbit. Because Kepler works by looking at such transit events it only sees planetary systems whose disk is aligned with Earth. If it were looking at earth from afar it would have only a one in 700 chance that this alignment occurred. If Kepler finds one

Earth-like planet we could guess that there are 700 in the sample they are looking at, which represents 1 millionth of the stars in our galaxy, but will it really find any?

My guess at this stage is that it will find a number of Earth-sized planets and a number of small planets in the habitable zone, but the statistics may be against it finding an Earth-sized planet in the habitable zone of a stable star like ours. Probably we will be able to estimate how many such planets there are and it may be something like a million in our galaxy. It could be a lot less. The next step will be tp determine how many are likely to have the right chemical mixture to form water and an Earth-like atmosphere. We don't yet know the answer, but it is exciting that the data we need to answer these questions is starting to become available.

References

1. http://blog.vixra.org/2011/12/22/searching-for-earths-twin/

Printed in Germany
by Amazon Distribution
GmbH, Leipzig